MIOCENE MAMMALS

AF875874

Books LLC®, Reference Series, Memphis, USA, 2011. ISBN: 9781150943645. www.
Copyright: http://creativecommons.org/licenses/by-sa/3.0/deed.en

Table of Contents

Introduction

Purchase of this book entitles you to a free trial membership in the publisher's book club at www.booksllc.net. (Time limited offer.) Simply enter the barcode number from the back cover onto the membership form. The book club entitles you to select from hundreds of thousands of books at no additional charge. You can also download a digital copy of this and related books to read on the go. Simply enter the title or subject onto the search form to find them.

Each chapter in this book ends with a URL to a hyperlinked online version. Type the URL exactly as it appears. If you change the URL's capitalization it won't work. Use the online version to access related pages, websites, footnotes, tables, color photos, updates. Click the version history tab to see the chapter's contributors. Click the edit link to suggest changes.

A large and diverse editor base collaboratively wrote the book, not a single author. After a long process of discussion and debate, the chapters gradually took on a neutral point of view reached through consensus. Additional editors expanded and contributed to chapters striving to achieve balance and comprehensive coverage. This reduced the regional or cultural bias found in many other books and provided access and breadth on subject matter otherwise little documented.

? *Nycticebus linglom*

? *Nycticebus linglom* is a fossil primate from the Miocene of Thailand. Known only from a single tooth, an upper third molar, it is thought to be related to the living slow lorises (genus *Nycticebus*), but the material is not sufficient to assign the species to *Nycticebus* with certainty, and the species name therefore uses open nomenclature. With a width of 1.82 mm, this tooth is very small for a primate. It is triangular in shape, supported by a single root, and shows three main cusps, in addition to various crests. The absence of a fourth cusp, the hypocone, distinguishes it from various other prosimian primates.

Taxonomy

? *Nycticebus linglom* was described in 1997 by French paleontologists Pierre Mein and Léonard Ginsburg in a report on the fossil mammals of Li Mae Long, a Miocene site in Thailand. The animal is known from a single tooth, and on the basis of comparisons with other prosimian primates Mein and Ginsburg concluded that it is most closely related to the living slow lorises (genus *Nycticebus*). However, in view of the very limited material, they only tentatively assigned the fossil species to *Nycticebus*, using open nomenclature. The specific name, *linglom*, is the Thai word for "loris".

Description

The single known tooth, a third upper molar (M3) known as T Li 41, is tiny, with a length of 1.29 mm and width of 1.82 mm. Mein and Ginsburg claim that it is the smallest known prosimian molar. The tooth is triangular in shape and shows a simple, reduced morphology. Three important cusps—the protocone, paracone, and metacone—are present, connected by a crest. They are low and rounded. The metacone, located at the back of the tooth, is closer to the protocone, which is on the front lingual corner (the side of the tongue), than it is to the paracone on the front labial corner (the side of the cheeks). The protocone is rounded on the lingual side and is attached to a weak crest on the front and back. On the front labial corner, a lengthy crest, the parastylar crest, is present, which includes a minor cusp known as a parastyle. Some wear is visible on the parastylar crest, and at the front of the tooth a contact facet with the preceding second upper molar is present. The tooth has a single, well-developed root, which contains a number of grooves, suggesting that it consists of three smaller, fused rootlets.

? *Nycticebus linglom* is much smaller than the fossil sivaladapine primates, and unlike tarsier M3s, the single known fossil is reduced in form and lacks a fourth main cusp, a hypocone. With its reduced, triangular form, it more closely resembles lorises (family Lorisidae), but the absence of a hypocone also distinguishes it from the slender lorises (*Loris*), the angwantibos (*Arctocebus*), and the pygmy slow loris (*Nycticebus pygmaeus*). The fossil genus *Nycticeboides* lacks the rounded lingual face of the protocone seen in ? *N. linglom* and possesses additional cuspules. However, ? *N. linglom* strongly resembles the Bengal slow loris (*Nycticebus bengalensis*), from which it is distinguished by its smaller size and fused roots.

Range and ecology

Li Mae Long, the collection site of ? *N. linglom*, is dated to the latest Early Miocene, corresponding to the European zone MN 4, around 18 mya. It is in the Thai province of Lamphun. The fossil fauna encompasses 34 species of mammals, including the tarsier *Tarsius thailandica* and the treeshrew *Tupaia miocenica*. Mein and Ginsburg conclude that the fauna represents a tropical forest environment close to a shallow lake.

Source (edited): "http://en.wikipedia.org/wiki/%3F_Nycticebus_linglom"

Aceratheriinae

Aceratheriinae is an extinct subfamily of rhinoceros endemic to Asia, Africa, Europe, and North America during the Oligocene through Pliocene living from 33.9—3.4 mya, existing for approximately 30.5 million years.

It consists of 6 genera:

- *Aceratherini* - Europe, Eurasia, Asia, Africa
- *Alicornops* - Europe, Eurasia, Asia
- *Aphelops* - North America
- *Chilotheriini* - Eurasia and Asia
- *Floridaceras* - North America
- *Peraceras* - North America

Taxonomy

Aceratheriinae was named by Dollo (1885). It was assigned to Rhinocerotida by Codrea (1992); and to Rhinocerotidae by Prothero (1998), Antoine et al. (2000), Kaya and Heissig (2001), Sach and Heizmann (2001) and Deng (2005).

Source (edited): "http://en.wikipedia.org/wiki/Aceratheriinae"

Aceratherium

Aceratherium was a genus of rhinoceros of the tribe Aceratheriini endemic to Africa and Asia from the Oligocene to Pliocene, living from 33.9—3.4 mya existing for approximately 30.5 million years.

Taxonomy

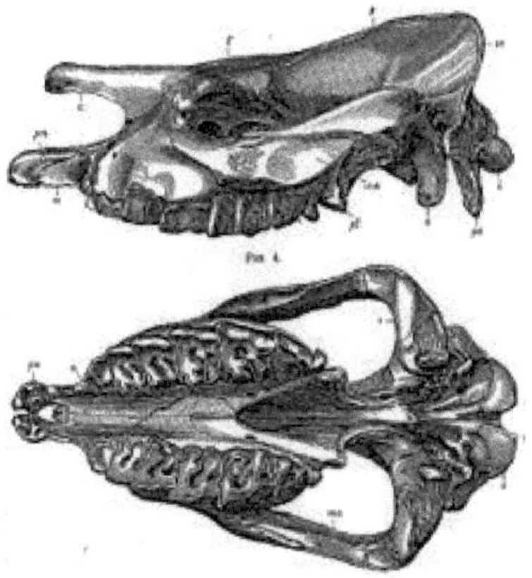

Skull

Restoration by Heinrich Harder

Aceratherium was named by Kaup (1832). It was assigned to Rhinocerotidae by Carroll (1988); and to Aceratherini by Kaya and Heissig (2001).
Source (edited): "http://en.wikipedia.org/wiki/Aceratherium"

Acerorhinus

Acerorhinus was a genus of rhinoceros of the tribe Aceratheriini endemic to Asia from the Miocene, living from 13.6—7.0 mya existing for approximately 6.6 million years.

Taxonomy

Acerorhinus was named by Kretzoi (1942). Its type is *Aceratherium zernowi*. It was assigned to Aceratherini by Kaya and Heissig (2001); and to Aceratheriini by Antoine and Saraç (2005).
Source (edited): "http://en.wikipedia.org/wiki/Acerorhinus"

Acrophyseter

Acrophyseter
Acrophyseter is a genus of stem-sperm whales that lived around 6 million years ago.

Description

Acrophyseter is derived from the Greek *akros*, meaning *acute*, which describes the short, pointed upturned snout, and *Physeter*, the scientific name for sperm whales. But unlike today's sperm whales, *A. deinodon* was relatively short at around 4 meters and had teeth on both the upper and lower jaws; 12 on each side of the upper jaw and 13 on each side of the lower. These teeth were large, sharp, and set firmly in deep sockets; the species epithet *deinodon* is from the Greek *deinos*, meaning terrible, and *odon*, tooth. The lower back teeth were close together suggesting they were used for shearing, unlike the suction feeding used by today's sperm whales, which don't even have teeth on their upper jaws. These differences imply that *Acrophyseter* fed on comparatively large prey, such as the smaller whales like *Piscolithax*, pinnipeds like *Acrophoca*, and penguins like *Spheniscus urbinai* that lived in the area at that time.

Various skull characteristics such as the size of the teeth and presence of tooth enamel which living sperm whales don't have indicate that *Acrophyseter* is probably a sister group to the *Brygmophyseter* and *Zygophyseter* group and to the group comprising *Aulophyseter* and living physeteroids, the kogiids and physeterids.
Source (edited): "http://en.wikipedia.org/wiki/Acrophyseter"

Adelphailurus

Adelphailurus is an extinct genus of saber-toothed cats of the family Felidae and tribe Metailurini which inhabited western North America during the Miocene living from 10.3—5.33 Ma and existed for approximately 4.97 million years.

Taxonomy

Adelphailurus was named by Hibbard (1934). Its type is *Adelphailurus kansensis*. It was assigned to Felidae by

Hibbard (1934) and Carroll (1988); and to Machairodontinae by Martin (1998).

Morphology

It was a cougar-sized animal and may have had habits similar to those of a cougar. Its body had the same shape as a cougar except for a long and compressed upper canine. This would place this cat into the "False-sabertooth" group. Apart from that the *Adelphailurus* had a retained upper second premolar, which is unusual for a cat.

Body mass

Two specimens were examined by Legendre and Roth for body mass.

- 9.56 kg (21 lb)
- 26.7 kg (59 lb)

Source (edited): "http://en.wikipedia.org/wiki/Adelphailurus"

Adilophontes

Adilophontes is an extinct genus of large, mostly carnivorous bone-crushing mammals known as bear dogs, of the family Amphicyonidae, endemic to North America during the Oligocene to Miocene, living from 24.8—20.6 Ma and existed for approximately 4.2 million years.

Taxonomy

Adilophontes was named by Hunt (2002). Its type is *Adilophontes brachykolos*. It was assigned to Daphoeninae by Hunt (2002).

Morphology

A single specimen was examined by Legendre and Roth for body mass and was estimated to weigh 68.6 kg (150 lb).

Fossil distribution

- Eighteen Mile District, Goshen County, Wyoming ~20.8 Ma.
- Lay Ranch Beds, Goshen County, Wyoming ~20.8 Ma.
- Guernsey, Platte County, Wyoming ~20.9 Ma.

Source (edited): "http://en.wikipedia.org/wiki/Adilophontes"

Adinotherium

Adinotherium (meaning "not *Deinotherium*") is an extinct genus of Toxodontidae, large bodied hoofed ungulates which inhabited South America during the Miocene living from 17.5—11.61 Ma and existed for approximately 5.89 million years.

Restoration

The 1.5 metres (4.9 ft) long animal looked like a smaller version of its rhinoceros-like relative *Toxodon*. Its front legs were somewhat longer than those of its relatives, making its hip and shoulder height about equal. A small horn atop *Adinotherium*'s skull may have played a role in the mating season.

Source (edited): "http://en.wikipedia.org/wiki/Adinotherium"

Aegyrcitherium

Aegyrcitherium was a genus of rhinoceros endemic to Europe during the Miocene living from 16.9—16 mya existing for approximately 0.9 million years.

Taxonomy

Aegyrcitherium was named by Antoine (1997). It was reranked as Hispanotherium (Aegyrcitherium) by Guérin and Pickford (2003). It was assigned to Elasmotheriini by Antoine et al. (2000) and Deng (2003); and to Hispanotherium by Guérin and Pickford (2003).

Source (edited): "http://en.wikipedia.org/wiki/Aegyrcitherium"

Aelurodon

Aelurodon is an extinct canine genus of the subfamily Borophaginae which lived from the Hemingfordian stage (20.6 Mya) of the Miocene to the Hemphillian stage of the Early Pliocene (4.9 Mya). *Aelurodon* existed for approximately 15.7 million years.

Origin

Aelurodon stirtoni

Aelurodon are a part of a clade of canids loosely known as "bone-crushing" or "hyena-like" dogs, that apparently descended from the earlier genera *Protomarctus* and *Tomarctus*. Several species are known from fossils found in the central and western U.S., suggesting a wide geographic range during their peak in the Miocene epoch. Large species of *Aelurodon* (*A. ferox* and *A. taxoides*) may have hunted in packs like modern wolves.

Canid competitors

Aelurodon lived during a period with other canidae, specifically Borophaginae such as *Epicyon* (20.6—5.330 Ma), *Paratomarctus* (16.3—5.3 Ma), *Borophagus* (23.3—3.6 Ma), *Carpocyon* (20.4—3.9 Ma), and the first emerging wolf, *Canis lepophagus* (10.3—1.8 Ma).

Species

Six species of *Aelurodon* are recognized. ***Aelurodon asthenostylus*** existed from 20.6—13.6 Mya (7 million years). It was named by Wang et al. in 1999. Fossils of *A. asthenostylus* have been uncovered in western Nebraska, northern Colorado, Nevada and south central California. Two specimens had an estimated body mass of 33.0 kg (73.5 lbs) and 29.4 kg (64.8 lbs). ***Aelurodon ferox*** existed from 16.3—10.3 Mya (6 million years) and was named by Leidy in 1858. Specimens have been found in western Nebraska, New Mexico, to the Texas gulf coast and to southern Montana. Two specimens had an estimated body mass of 45.6 kg (100.5 lbs) and 39.9 kg (88 lbs). ***Aelurodon mcgrewi*** existed from 16.3—13.6 Mya (3 million years) with fossils known from western Nebraska. Two specimens had an estimated body mass of 30.6 kg (67.4 lbs) and 34.7 kg (76.5 lbs). ***Aelurodon montaneis*** existed from 16.3—13.6 Mya (3 million years). ***Aelurodon stirtoni*** existed from 16.3—10.3 Ma (6 million years). Fossil specimens were uncovered in western Nebraska and New Mexico. Two specimens had an estimated body mass of 33.9 kg (74.7 lbs) and 29.3 kg (64.5 lbs). ***Aelurodon taxoides*** existed from 13.6—5.330 Mya (8.27 million years). Fossil specimens were uncovered in Florida, south and north Texas, New Mexico, Kansas, and the bay area of California. Two specimens had an estimated body mass of 56.3 kg (124 lbs) and 48.8 kg (107.5 lbs). The name *Aelurodon* was given by Joseph Leidy in 1858. The smaller *Aelurodon* lineage evolved teeth adapted to a more purely carnivorous (hypercarnivorous) diet, a trend consistent with other borophagines. Some specimens of the largest species, *Aelurodon taxoides*, reached the size of a tiger.

Source (edited): "http://en.wikipedia.org/wiki/Aelurodon"

Aeluroidea

Aeluroidea is an extant and extinct superfamily of feline-like carnivores which are or were endemic to North America, South America, Africa, and Asia. They appeared during the Oligocene about 33.3 million years ago.

This superfamily includes extant families of: Felidae (cats), Herpestidae (mongoose), Hyaenidae, Nandinia (civet) and the extinct *Africanictis, Anictis, Asiavorator, Haplogale, Herpestides, Mioprionodon, Moghradictis, Palaeoprionodon, Proailurus, Shandgolictis, Stenogale*, and *Stenoplesictis*.

Taxonomy

Aeluroidea was named by Flower (1869). It is extant. It was assigned to Carnivora by Flower (1883) and Carroll (1988); and to Feliformia by Bryant (1991).

Source (edited): "http://en.wikipedia.org/wiki/Aeluroidea"

Aepycamelus

Aepycamelus is an extinct genus of camelid, formerly called ***Alticamelus*** which lived during the Miocene 20.6-4.9 Ma existing for approximately 15.7 million years.

Its name is derived from the Homeric Greek αἰπύς, "high and steep" and κάμελος - "camel"; thus, "high camel"; *alticamelus* in Latin. *Aepycamelus* walked on its toes only. Unlike earlier species of camelid, they possessed cushioned pads like those of modern camels.

Morphology

Aepycamelus was a prairie dweller of North America (Colorado, etc). It was a highly specialized animal. Its head was relatively small compared with the rest of its body, its neck was long, as a result of giraffe-like lengthening of the cervical vertebrae, and its legs were long and stilt-like, with the elbow and knee joints on the same level. The top of its head would have been about 3 metres (9.8 ft) above the ground.

Body mass

Four fossil specimens were examined to determine body mass by M. Mendoza,

C. M. Janis, and P. Palmqvist. The findings are:

- Specimen 1: 721.3 kg (1,600 lb)
- Specimen 2: 621 kg (1,400 lb)
- Specimen 3: 574.7 kg (1,300 lb)
- Specimen 4: 363.5 kg (800 lb)

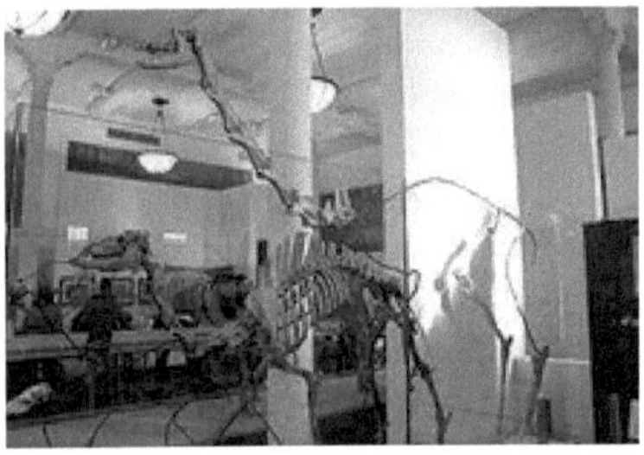

Aepycamelus giraffinus and *Procamelus grandis*

Its strange body structure gives us plenty of information on its mode of life and habits. *Aepycamelus* obviously inhabited dry grasslands with groups of trees. It is presumed to have moved about singly or in small groups, like today's giraffes, and like them, browsed high up in the trees. In this respect it had no competitors, it survived a relatively long time, lasting throughout the entirety of the Miocene epoch, and died out prior to the start of the Pliocene, possibly due to climatic changes.

Fossil distribution

Fossil distribution is widespread from Montana to Florida to California.

Source (edited): "http://en.wikipedia.org/wiki/Aepycamelus"

Afrocyon

Afrocyon is an extinct genus of large, mostly carnivorous bone-crushing mammals known as bear dogs, of the family Amphicyonidae endemic to Africa during the Miocene to Pliocene living from 11.6—5.3 Ma and existed for approximately 6.3 million years.

Taxonomy

Afrocyon was named by Arambourg (1961). It was assigned to Amphicyonidae by Carroll (1988).

Source (edited): "http://en.wikipedia.org/wiki/Afrocyon"

Afropithecus

Afropithecus turkanensis, a new Miocene hominoid, was excavated from a small site near Lake Turkana called Kalodirr in northern Kenya in 1986 and was named by Richard Leakey and Meave Leakey. The estimated age of *Afropithecus* is between 16 and 18 million years old, which was determined with radiometric dating techniques and the geological studies conducted by Broschetto and Brown from the University of Utah. In total there are 46 recovered specimens from Kalodirr relating to *Afropithecus* consisting of cranial, mandible, dentition and post-cranial remains. The type specimen of *Afropithecus turkanensis* is KNM-WK 16999.

Morphology

Leakey and Leakey first described *Afropithecus turkanensis* to be a large hominoid which appeared to have relatively thick enamel. Leakey suggested that *A. turkanensis* shared postcranial features with the species *Prosconsul nyanzae*, which is the best known genus Miocene with literally hundreds of fossils having been found representing almost all skeletal elements, and sharing cranial features with *Aegyptopithecus zeuxis* and dental features with *Heliopithecus leakeyi* which had two weathered molars that indicated a general distinction from known large early catarrhines, and later concluded that *A. turkenensis* was a primitive, arboreal quadruped similar to *P. nyanzae*, and that *A. turkanensis* had primitive facial morphology and derived dental characteristics that would suggest a diet of hard fruits.

Cranial Morphology

The type specimen, KNM-WT 16999 is composed of a long distinct snout, the facial skeleton, frontal, much of the coronal structure, most of the sphenoid, and relatively unworn adult dentition; the right orbit (virtually complete), the right zygomatic, the pterygoid, most of the sphenoid and lesser wings, the maxilla and premaxilla, and adult dentition with procumbent incisors. The surface on the right side maxilla and premaxilla along with the enamel on the right molars has been lost over time and has been replaced with calcite crystals, which only provide the general shape and not the details.

From dentition its known that the palate, which is almost completely calcified, of *A. turkanensis* is shallow, long and narrow with tooth rows that converge posteriorly, and it is probable the tooth rows were originally nearly parallel. A. turkanensis had a 6.5mm diastema between its very procumbent second incisor (KNM-WT 16999 had large, broad incisors) and the canine.

The thickness of the enamel on the molars is often reported when fossils are being recorded and used to make comparisons across taxa. The thickness is referred to either as "thin" or "thick" and is commonly assessed as a linear measurement of the enamel on worn or naturally fractured teeth. From enamel testing it has been suggested that *A. turkanensis* is the oldest known thick-enamelled hominoid, which is what would distinguish it from *Kenyapithe-*

cus.

Post-Cranial Morphology

Post cranial remains such as KNM-WK 16901, includes an associated right fibula (lacking the proximal portion, and is approximately the same size as *Pan troglodytes*; 184 mm), a right proximal third metatarsal, a right fourth metatarsal lacking the head, and an incomplete first metatarsal head.

Other post-cranial remains include: KNM-WK 17016P a large right ulna, and foot or hand bones: KNM-WK 17008, KNM-WK 18395.

Source (edited): "http://en.wikipedia.org/wiki/Afropithecus"

Agnotherium

Agnotherium is an extinct genus of large terrestrial carnivores belonging to the suborder Caniformia, family Amphicyonidae ("bear dog"), and which inhabited Euroasia and North Africa from the Early Miocene subepoch to the Middle Miocene subepoch 16.9—11.6 Mya, existing for approximately 5.3 million years.

Agnotherium was originally assigned to Thaumastocyonini by Ginsburg in 1977 and to Amphicyonidae by Carroll in 1988.

Fossil distribution

The first specimen was located in strata zone MN 4 in Alsace, France Other locations were: En Pejouan, Midi-Pyrenees Region; Pontigne. Another specimen was discovered in Bèni Mellal, Morocco.

Species

- *Agnotherium antiquus*
- *Agnotherium grivense*

Source (edited): "http://en.wikipedia.org/wiki/Agnotherium"

Agriotherium

Agriotherium is an extinct genus of Ursidae of the Miocene through Pleistocene epochs, endemic to North America, Europe, Africa, and Asia living from ~13.6–2.5 Ma, existing for approximately 11.1 million years.

Taxonomy

Agriotherium was named by Wagner (1837). It was assigned to Agriotheriini by Chorn and Hoffman (1978); to Hemicyoninae by Qiu et al. (1991); to Ursavini by Hunt (1998); to Ursidae by Wagner (1837), Carroll (1988) and Salesa et al. (2006), and Ursinae by Krause et al. 2008.

Morphology

Agriotherium was about 2.7 metres (9 ft) in body length, making it larger than most living bears. Except for the extinct subspecies of modern polar bear *Ursus maritimus tyrannus*, *Agriotherium* was along with the short-faced bear, *Arctodus simus* the largest member of terrestial Carnivora. It had dog-like crushing teeth. Its primary diet was carnivorous and secondary was omnivorous possibly classifying this animal as mesocarnivore. With a body mass greater than most large ungulates (horses, bovines, camelids, and rhinoceroses), it is probable that *Agrotherium* could have preyed on these.

Body mass

Two specimens were examined by Legendre and Roth for body mass.

- Specimen 1: 79.3 kg (170 lb)
- Specimen 2: 652.6 kg (1,400 lb)

Fossil distribution

Sites and age of specimens:

- American Cyanimid Company site, Bone Valley Formation, Polk County, Florida ~13.7—11.6 Ma.
- Venta del Moro, Spain ~9–5.3 Ma.
- Lang. E Quarry, South Africa ~5.3–3.6 Ma.
- Carlin High Quarry, Elko County, Nevada ~23–5.3 Ma.
- Vialette, Haute Loire, France ~3.2–2.5 Ma.
- Middle Awash, Ethiopia ~11.6–3.6 Ma.

Agriotherium ranged widely; fossils of four or more species have been found in Europe, India, China, North America and South Africa. It is the only ursoid known to have colonized sub-Saharan Africa (amphicyonid "bear dogs" also reached the area).

References

Further reading

- Dalquest, W. W. (1986). "Lower Jaw and Dentition of the Hemphillian Bear, Agriotherium (Ursidae), with the Description of a New Species". *Journal of Mammalogy* **67** (4): 623–631. doi:10.2307/1381124. JSTOR 1381124.
- Miller, W. E.; Carranza-Castaeda, O.; Carranza-Castaneda, Oscar (1996). "Agriotherium schneideri from the Hemphillian of Central Mexico". *Journal of Mammalogy* **77** (2): 568–577. doi:10.2307/1382830. JSTOR 1382830.
- Petter, G.; Thomas, H. (1986). "Les Agriotheriinae (Mammalia, Carnivora)néogènes de l'Ancien Monde presence du genre Indarctos dans la faune de Menacer (ex–Marceau), Algérie". *Geobios* **19**: 573–586.
- Sorkin, B. (2006). "Ecomorphology of the giant short-faced bears Agriotherium and Arctodus". *Historical Biology: A Journal of Paleobiology* **18**: 1–20. doi:10.1080/08912960500476366.

Source (edited): "http://en.wikipedia.org/wiki/Agriotherium"

Aguascalientia

Aguascalientia is an extinct genus of terrestrial herbivore the family Camelidae, endemic to North America during the Early Miocene 23.030—20.6 mya existing for approximately 2.43 million years.

Taxonomy

Aguascalientia was named by Stevens (1977). It was assigned to Camelidae by Stevens (1977) and Carroll (1988).

Morphology

Fossil specimens of one individuals' body mass were examined by M. Mendoza, C. M. Janis, and P. Palmqvist. The specimen was estimated to weigh 34.1 kg (75 lb). The second specimen was estimated to weigh 5.52 kg (12 lb).
Source (edited): "http://en.wikipedia.org/wiki/Aguascalientia"

Ailurarctos

***Ailurarctos** (panda bear)* is an extinct genus of panda from the Late Miocene of China, some 8 million years ago.
Source (edited): "http://en.wikipedia.org/wiki/Ailurarctos"

Aivukus

Aivukus is an extinct genus of walrus from the Miocene. From fossil records it was a much bigger, but had a walrus skeleton, but much enlarged. It probably ate fish and plants. The behavior was quite different, as fossil hunters have uncovered two skeletons with large chunks out of the backs of their tusks, which was too hard to have been a fight between the two, it was quite like a rock had scratched it, or they used their tusks to mark their territory on rocks.
Source (edited): "http://en.wikipedia.org/wiki/Aivukus"

Aletomeryx

Aletomeryx is an extinct genus of Artiodactyla, of the family Palaeomerycidae, endemic to North America from the early Miocene epoch (Hemingfordian stage) 20.6—16.3 Ma, existing for approximately 4.3 million years.

Taxonomy

Aletomeryx was named by Lull (1920). It is the type genus of Aletomerycinae, Aletomerycini. It was assigned to Dromomerycidae by Lull (1920) and Janis and Manning (1998); and to **Aletomerycini** by Prothero and Liter (2007).

Fossil distribution

Fossils have been recovered from the Midway Site in Florida, Saskatchewan, Boron, California, and several sites in Nebraska and Wyoming.
Source (edited): "http://en.wikipedia.org/wiki/Aletomeryx"

Alforjas

Alforjas is an extinct genus of terrestrial herbivore the family Camelidae, endemic to North America during the Miocene through Pliocene 10.30—5.3 mya existing for approximately 5 million years.

Taxonomy

Alforjas was named by Harrison (1979). Its type is Alforjas taylori. It was assigned to Camelidae by Harrison (1979) and Carroll (1988).

Morphology

A single specimen was examined for estimated body mass by M. Mendoza, C. M. Janis, and P. Palmqvist. The specimen was estimated to weigh 623.3 kg (1,400 lb).
Source (edited): "http://en.wikipedia.org/wiki/Alforjas"

Alkwertatherium

Alkwertatherium is an extinct genus of marsupial of the family Diprotodontidae. Only one species was descript, ***Alkwertatherium webbi***, from Late Miocene of Northern Territory.
Source (edited): "http://en.wikipedia.

org/wiki/Alkwertatherium"

Allodesmus

Allodesmus is an extinct genus of pinniped from the Miocene and related to the genus *Desmatophoca*. It measured about 8 feet (2.4 m) long and weighed 800 pounds (360 kg).
Source (edited): "http://en.wikipedia.org/wiki/Allodesmus"

Alopecocyon

Alopecocyon was a small relative of the modern Red Panda. It weighed only about 11 lbs (5 kg). It was closely related to Simocyon, a larger member of its group. Its fossils have been found in France, Poland, and Slovakia.
Source (edited): "http://en.wikipedia.org/wiki/Alopecocyon"

Amebelodon

Amebelodon is a member of a diverse group of primitive proboscideans called gomphotheres, a group that also gave rise to the modern elephants and their close relative the mammoth. The most striking attribute of this animal is its lower tusks, which are narrow, elongated,and distinctly flattened with the degree of flattening varying among the different species. *Amebelodon* first appeared in the Great Plains and Gulf Coast regions of North America during the late Miocene, roughly between 9 and 8 million years ago, and apparently became extinct on this continent sometime around 6 million years ago. It managed to migrate to Asia via the Bering land bridge where it has been found in a number of late Miocene sites, particularly in China. The youngest record of *Amebelodon* is from a 5 million year old site in North Africa.

Species

Gomphotheres are notoriously difficult to identify with confidence down to the species level because of considerable intraspecific variation in the group, with a result that many named species are probably invalid. However, reasonably well established *Amebelodon* species include:

- *A. britti*
- *A. cyrenaicus*
- *A. floridanus*
- *A. fricki*

Amebelodon varied considerably in size. *Amebelodon floridanus* was smaller than a modern Asian elephant, while *Amebelodon britti*, reached a shoulder height of 8 to 10 feet at the shoulder and a likely weight of 10 metric tons, similar to the largest mammoths and considerably larger than the living African elephant.

Paleobiology

Like other typical gomphotheres, *Amebelodon* possessed two sets of tusks, two uppers (much like those found on modern elephants), and two lowers that extended from the very front of the lower jaws. However, as mentioned above the lower tusks of *Amebelodon* were distinctive in that they were relatively long, slender, and flattened. Because of the resemblance of these lower tusks to shovels, *Amebelodon* is commonly referred to as a "shovel-tusked" gomphothere (another shovel-tusked gomphothere that may or may not be closely related to *Amebelodon* is *Platybelodon*). There has long been an assumption that these lower tusks were actually used as shovels by the animal during feeding, presumably to scoop up water plants. However, an analysis of wear patterns has shown that these lower tusks were most likely used in a variety of ways in addition to shoveling, including scraping bark from trees. Overall, the evidence indicates that this animal was a versatile browser (an animal that eats broad-leaved plants rather than grass), feeding in both wet and dry settings in a variety of ways. For reasons that are not clearly explained in the literature, *Amebelodon* is usually portrayed with a short, flap-like trunk (this is true of the picture in the external link below). However, considerable evidence indicates that *Amebelodon* actually had a long, flexible trunk very much like that found in modern elephants, and that it probably was important for feeding.
Source (edited): "http://en.wikipedia.org/wiki/Amebelodon"

Ampelomeryx

Ampelomeryx is an extinct genus of herbivorous even-toed ungulate mammal belonging to the family Palaeomerycidae

Ampelomeryx was named by Duranthon et al. (1995). It was assigned to *Palaeomerycinae* by Prothero and Liter (2007). It had frontal and occipital appendages. It was similar to *Tauromeryx* and *Triceromeryx*.
Source (edited): "http://en.wikipedia.org/wiki/Ampelomeryx"

Amphechinus

Amphechinus is an extinct genus of hedgehog of the family *Erinaceidae*, endemic to North America, Africa, Europe, and Asia during the Pleistocene epoch (23.8—1.8 mya), existing for approximately 22 million years.

Taxonomy

Amphechinus was named by Aymard (1850). It is not extant. It was assigned to *Erinaceidae* by Aymard (1850), Rich (1981) and Pickford et al. (2000).

Morphology

A single specimen was examined by J. I. Bloch, K. D. Rose, and P. D. Gingerich for body mass and was estimated to have a weight of 175 g (0.39 lb).

Its fossils have been found in Africa, Asia, Europe, and North America.

Species

A. akespensis, A. arvernensis, A. baudelotae, A. edwardsi, A. ginsburgi, A. golpeae, A. horncloudi, A. intermedius, A. kreuzae, A. major, A. microdus, A. minutissimus, A. robinsoni, A. taatsiingolensis.

Source (edited): "http://en.wikipedia.org/wiki/Amphechinus"

Amphicyon

Amphicyon ("ambiguous dog") is an extinct genus of large carnivorous bone-crushing mammals, known as bear-dogs, of the family Amphicyonidae, subfamily Amphicyoninae, from the Aquitanian Epoch until the Tortonian. They ranged over North America, Europe, Asia, and Africa from 20.6—9 Ma ago, existing approximately 11.6 million years.

Morphology

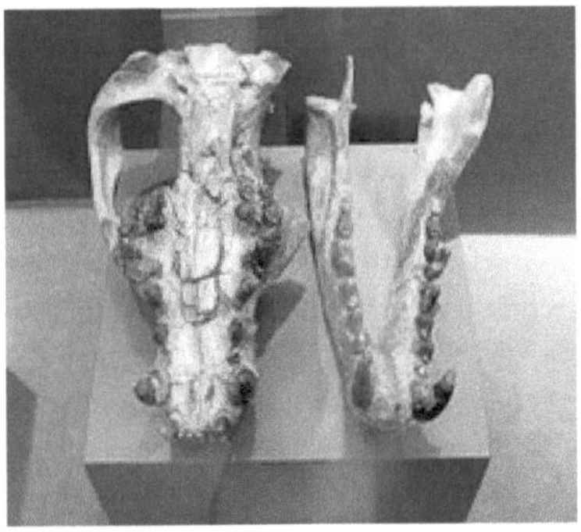

Jaws, Paläontologische Museum München

Amphicyon was the typical bear-dog amphicyonid with morphology similar to both bears and dogs. With its robust build and maximum length of 2.5 m (8 ft), the largest species looked more like a bear than a dog. It had a large heavy tail, thick neck, robust limbs and teeth like a wolf. It was probably an omnivore with a lifestyle comparable to that of the brown bear.

Body mass

A single specimen was examined by Legendre and Roth and estimated to have a body mass of 84.2 kg (190 lb), roughly half that of *Ischyrocyon* and twice that of *Epicyon* which shared its time period and habitat. *A. ingens* was much bigger: Sorkin (2008) estimated the largest known specimen (AM 68108) to weigh 600 kg, making it the largest amphicyonid and one of the largest known carnivorous land mammals.

Fossil distribution

The earliest occurrences of *Amphicyon* in North America are from the early to mid-Miocene, found in the Runningwater Formation in Sioux County, Nebraska, and from the lower part of the Troublesome Formation, Colorado (*A. galushai, A. frendens*, and *A. ingens*). Although other large amphicyonids from the Miocene of North America have been placed in *Amphicyon*, many of these carnivores are now placed in other amphicyonid genera. The *Amphicyon* lineage in the New World is restricted to the above three species (18.8–14.2 Ma). Particularly rich samples of the large North American species of *Amphicyon* have been found in the Sheep Creek Formation (*A. frendens*) and Olcott Formation (*A. ingens*) of central Sioux County, northwest Nebraska. *Amphicyon* has also been found in France, Spain and Germany in Europe.

Species

Amphicyon frendens lived from 20.16–13.6 Ma, approximately 6.56 million years. The species was originally described by W. Matthew in 1924 from specimens found in the middle member of the Sheep Creek Formation, Sioux County, Nebraska. *A. frendens* specimens have since been found at sites in Harney and Malheur Counties, Oregon. A specimen examined by S. Legendre and C. Roth in 1988 yielded an estimated body mass of 135.6 kg (300 lb), similar to that of *Ischyrocyon*, *Amphicyon galushai* and its borophagine competitor *Epicyon*, which it coexisted with.

Amphicyon galushai lived from 20.6–16.3 Ma, approximately 4.3 million years. *A. galushai* was described by R. Hunt in 2003 based on the type specimen F:AM 25406, found in the Dunlap Camel Quarry, Runningwater Formation, Dawes County, Nebraska Specimens assigned to this species have since been found at several sites in Nebraska and one site in Colorado.

Amphicyon giganteus lived from 16.8–7.2 Ma, approximately 9.7 million years, in what is now Europe and Namibia. The species was first described in 1884 by Kaup. A specimen of *Iberotherium rexmanueli zbyszewskii* with teeth marks from *A. giganteus* was found in Portugal. It is unknown if he young *Iberotherium* was attacked or the carcass found and scavenged. The find was described by paleontologists Antunesa et al. in 2006.

Amphicyon ingens lived from 20.06–13.6 Ma, approximately 6.46 million years. The species was originally described by W. Matthew in 1924 from specimens found in the Olcott Formation, Sioux County, Nebraska. Specimens attributed to this species have since been found in California, Colorado, and New Mexico.

Amphicyon major lived from 16.9–9.0 Ma, approximately 7.9 million years. Specimens have been found in across Europe and in western Turkey. The species was named by De Blainville in 1841.

Source (edited): "http://en.wikipedia.org/wiki/Amphicyon"

Amphicyoninae

Amphicyoninae is a subfamily of extinct Amphicyonidae, large terrestrial carnivores (bear-dogs) belonging to the suborder Caniformia (meaning "dog-like") and which inhabited North America, Europe, Eurasia, Asia. and Africa from ~22.4—4.9 Ma. *Amphicyoninae* existed for approximately ~17.5 million years.

Amphicyoninae was named by Trouessart (1885). It was assigned to Canidae by Matthew (1902); to Ursidae by Ginsburg (1977); and to Amphicyonidae by Hunt (1998).

Members include:

- Amphicyon, endemic to Europe and N. America
- Cynelos, synonymous with Absonodaphoenus and Hecubides, endemic to N. America
- Ischyrocyon, synonymous with Hadrocyon, endemic to N. America
- Pliocyon, endemic to N. America
- Pseudocyon, synonymous with Amphicyonopsis, endemic to Europe and N. America
- Thaumastocyonini, a tribe endemic to Europe
- Ysengrinia, endemic to Europe and N. America

Fossil distribution

Specimens have been recovered from:

- Midway Site, Gadsden County, Florida ~18.9—18.8 Ma.
- Rabbitt Creek Site, Meagher County, Montana ~21.7—18 Ma.
- Grimes Landing Site, King William County, Virginia ~17.6 Ma.
- Arrisdrift, Namibia ~23.03—11.6 Ma.
- Les Beilleaux, France ~20—16.9 Ma.

Source (edited): "http://en.wikipedia.org/wiki/Amphicyoninae"

Anancus

Anancus is an extinct genus of gomphothere endemic to Africa, Europe, and Asia, that lived during the Turolian age of the late Miocene to early Pleistocene, roughly from 3—1.5 million years ago.

Anancus stood around 3 metres (9.8 ft) tall, and closely resembled a modern elephant. It had two tusks, whereas most other gomphotheres had four. Aside from its somewhat shorter legs, *Anancus* was also different from modern elephants in that its tusks were much longer, up to 4 m (13 ft) in length. The tusks were possibly defense weapons not unlike elephants of today. The molars were not composed of lamellae like those of true elephants, but had cusps, like tapir and pig molars; *Anancus* appears to have lived in forests, eating from trees and shrubs and digging out tubers and roots in the forest floor, and it died out when these forests gave way to grasslands.

Although not as famous, *Anancus* was at least as big as its cousins the mammoths.

Taxononmy

Anancus was named by Aymard in 1855. This animal's species type is *Anancus arvernensis*. *Anacus* was assigned to the suborder *Elephantiformes* by Jeheskel Shoshani, Ph.D. and P. Tassy in 1996. It was then assigned to the family *Gomphotheriidae* by Carroll in 1988. It was assigned to the family *Elephantidae* by McKenna and Bell in 1997 and by Lambert and Shoshani in 1998. *Anancus* was then assigned to the superfamily of *Elephantoidea* by Kalb and Froelich in 1995 and Shoshani and Tassy in 2005. In 2009, L. Hautier, H. T. Mackaye, F. Lihoreau, P. Tassy, P. Vignaud and M. Brunet assigned it collectively to the subfamily *Anancinae*.

Source (edited): "http://en.wikipedia.org/wiki/Anancus"

Ancylotherium

Ancylotherium (from Greek, meaning "hooked beast") is an extinct genus of the family Chalicotheriidae, subfamily Schizotheriinae, endemic to Europe, Asia, and Africa during the Late Miocene-Pliocene (9.0—1.8 mya), existing for approximately 7.2 million years.

Taxonomy

Ancylotherium was named by Gaudry (1863) and was assigned to Chalicotheriidae by Carroll (1988); and to Schizotheriinae by Geraads et al. (2007).

Morphology

At two meters high at the shoulder, *Ancylotherium* was relatively large, and was built rather like a goat. Like other chalicotheres, it had long forelimbs and short hind limbs. However, unlike most other chalicotheres, it did not walk on its knuckles. It was similar to the North American genus *Moropus*.

Environment

Ancylotherium's habitat was the savannahs of East and South Africa. As an herbivore, it evolved to browse on vegetation on the trees in the grassy savannahs of Africa. *Ancylotherium's* closest relatives are the other perissodactyls, or "odd-toed" ungulates, including the extinct brontotheres and modern-day mammals such as horses, tapirs, and rhinoceroses.

Fossil distribution

Fossil remains of *Ancylotherium* have been found at many of the hominid fossil sites in East and South Africa, including sites in Laetoli, Olduvai and Omo.

In popular culture

Ancylotherium model (middle) used in Walking with Beasts on exhibit at Horniman Musem, London

Ancylotherium appears in the BBC's series *Walking with Beasts*, where CG animation was used to recreate extinct creatures of the Cenozoic era.
Source (edited): "http://en.wikipedia.org/wiki/Ancylotherium"

Andegameryx

Andegameryx is an extinct genus belonging to the family Hypertragulidae, within the order Artiodactyla, endemic to Europe during the Miocene, living 22.4—20 Ma, existing for approximately 2.4 million years.

Andegameryx were primitive and ancient ruminants, resembling small deer or musk deer, although they were more closely related to modern chevrotains. It's diet is stated to be that of a frugivore. The only known fossil was found in Zaragoza, Spain.

Taxonomy

Andegameryx was named by Ginsburg (1971). It was assigned to Hypertragulidae by Carroll (1988).
Source (edited): "http://en.wikipedia.org/wiki/Andegameryx"

Anoiapithecus

Anoiapithecus is an extinct genus of ape thought to be closely related to *Dryopithecus*. Both genera lived during the Miocene, approximately 12 million years ago. Fossil specimens named by Salvador Moyà-Solà *Anoiapithecus brevirostris* are known from deposits from Spain.

The discoverers describe *Anoiapithecus brevirostris* as a hominoid (superfamily Hominoidea) in the Dryopithecine tribe. They believe that it has more modern traits then the Kenyapithecines from which Kenyapithecus wickeri from Kenya brings fragmentary information. The African specimens are considered a sister taxon to the hominids, and 2 million year younger European specimens must be from after the time that these two groups split. This means that hominids may have evolved in Europe.

The name comes from the Anoia River region in Catalonia, where the fossil was found. It has been given the nickname *Lluc* (since it is a male individual). The name Lluc is the Catalan form of Luke, which in Latin suggests "light" and this discovery enlighted our early evolution

The modern anatomical features that characterized the family Hominidae visible in Lluc's fossil among others are: unique facial pattern for hominoids, nasal aperture wide at the base, high cheek bone, deep palate.
Source (edited): "http://en.wikipedia.org/wiki/Anoiapithecus"

Anthracotherium

Anthracotherium ("Coal Beast") was a genus of extinct artiodactyl ungulate mammals, characterized by having 44 teeth, with five semi-crescentic cusps on the crowns of the upper molars. The genus ranged throughout the Oligocene period, having a distribution throughout Europe, Asia, and North America. They died out during the mid to late Miocene, possibly due to a combination of climatic change and competition from other artiodactyls, including pigs and hippopotami (such may have been the case in Europe).

The genus typifies the family Anthracotheriidae, if only because it is the most thoroughly studied. In many respects, especially the anatomy of the lower jaw, *Anthracotherium*, as with the other members of the family, is allied to the hippopotamus, of which it is probably an ancestral form. Recent evidence further suggests that anthracotheres, together with hippos, may be close to the ancestry of the whales.

The genus name stems from the fact that the remains first described having been obtained from the Tertiary lignite-beds of Europe.

The European *Anthracotherium magnum* was approximately as large as a hippo, but there were several smaller species and the genus also occurs in Egypt, India and North America. Members of the genus *Anthracotherium*, as well as other members of the family Anthracotheriidae, are known colloquially as anthracotheres.

Anthracotherium monsvialense mandible

Anthracotherium sp. coprolite at the Museum für Naturkunde, Berlin

Source (edited): "http://en.wikipedia.org/wiki/Anthracotherium"

Aphelops

Aphelops is an extinct genus of rhinoceros endemic to North America during the Miocene through the Pliocene, living from 20.43—5.330 mya, existing for approximately 15.1 million years.

Taxonomy

Aphelops was named by Owen (1845) with species attributed to Cope 1873 and Cope 1874. It was assigned to Aceratheriinae by Prothero (1998); and to Rhinocerotidae by Owen (1845), Carroll (1988) and Hulbert and Whitmore (2006).

Morphology

Body mass

Five specimens were examined by M. Mendoza, C. M. Janis, and P. Palmqvist. The results were:

- Specimen 1: 265.7 kg (590 lb)
- Specimen 2: 474.0 kg (1,000 lb)
- Specimen 3: 3,327.5 kg (7,300 lb)
- Specimen 4: 1,283.3 kg (2,800 lb)
- Specimen 5: 261.9 kg (580 lb)

Source (edited): "http://en.wikipedia.org/wiki/Aphelops"

Archaeobelodon

Archaeobelodon is an extinct genus of proboscidea of the family *Gomphotheriidae* endemic to Europe and Africa during the Miocene from 16.9—16.0 Ma, living for approximately 0.9 million years.

Archaeobelodon was an ancestor of *Platybelodon* and *Amebelodon*. *Archaeobelodon* had a trunk and tusks.

Source (edited): "http://en.wikipedia.org/wiki/Archaeobelodon"

Archaeopotamus

Archaeopotamus is an extinct genus of Hippopotamidae that lived between 7.5 and 1.8 million years ago in Africa and the Middle East. The genus was described in 2005 to encompass species of hippos that were previously grouped in *Hexaprotodon*.

Archaeopotamus means "the ancient of the river". Of all identified hippos, only *Kenyapotamus* is older. *Kenyapotamus*, however, is only known from partial fossils; *Archaeopotamus* is the oldest well-identified hippo.

Characteristics

Many prehistoric hippo fossils are known primarily through fossils of the lower jaw. *Archaeopotamus*, like *Hexaprotodon*, has three pairs of incisors. Unlike other *Hexaprotodon*, *Archaeopotamus* has a highly elongate mandibular symphysis. The informal name "narrow muzzled hippos" has been suggested for this genus.

A. lothagamensis

A. lothagamensis is a species whose fossils were discovered at Lothagam, a site southwest of Lake Turkana in Kenya and first described in 2000. When first described, the species was classified as *Hexaprotodon lothagamensis*, although the significant morphological differences between other *Hexaprotodon* species, were noted.

A. lothagamensis was smaller than the modern hippopotamus and also smaller than *A. harvardi* and with a more slender skeleton. It was, however, significantly larger than the modern pygmy hippopotamus.

A. harvardi

A. harvardi is a species of hippo first described in 1977, originally as

Hexaprotodon harvardi. Although the proportions of *A. harvardi* and *A. lothagamensis* are similar, the former species is significantly smaller. Femurs of *A. harvardi* are approximately the same size as those of the modern hippopotamus.

Another group of fossils, originally described as *Hexaprotodon sahabiensis* or the Abu Dhabi Hippopotamus, are now considered to belong to *A. harvardi* or *A. lothagamensis*. The fossil record for *A. harvardi* is more extensive than for other *Archaeopotamus*.

A. harvardi is believed to have had a more riparian lifestyle than *A. lothagamensis*.

Distribution

Fossils of *Archaeopotamus* have been unearthed near Lake Turkana, Kenya; near Lake Victoria in Kenya and in Abu Dhabi and thus likely ranged across East Africa and the Arabian Peninsula.

Evolution

The immediate ancestor of *Archaeopotamus* is unknown. Whether or not *Archaeopotamus* is descended from *Kenyapotamus* cannot be determined from the few fossils available. Although *Archaeopotamus* is more primitive than any member of the genus *Hippopotamus*, it is likely a sister group of both hippopotamuses and *Hexaprotodon*.

Fossils similar to *Archaeopotamus* have been dated to as recently as about 2 million years ago, the end of the Pliocene epoch. These fossils are significantly different from those of *Hippopotamus* and *Hexaprotodon* of the same time.

Source (edited): "http://en.wikipedia.org/wiki/Archaeopotamus"

Arctoidea

Arctoidea is a superfamily of extinct and extant mostly carnivorous mammals which include the extinct group Hemicyonidae (dog-bears), and extant groups Musteloidea (weasels), *Nothocyon*, Pinnipedia (seals, sea lions), and Ursidae (bears), found in all continents from the Eocene, 46 Ma ago, to the present, approximately 46 million years..

Taxonomy

Arctoidea was named by Flower (1869). It was reranked as the unranked clade Arctoidea by Hunt (2001), Hunt (2002) and Hunt (2002); it was reranked as the infraorder Arctoidea by Koretsky (2001), Zhai et al. (2003) and Labs Hochstein (2007). It was assigned to Carnivora by Flower (1883), Barnes (1987), Barnes (1988), Carroll (1988), Barnes (1989), Barnes (1992), Hunt (2001), Hunt (2002) and Hunt (2002); and to Caniformia by Tedford (1976), Bryant (1991), Wang and Tedford (1992), Tedford et al. (1994), Koretsky (2001), Zhai et al. (2003), Wang et al. (2005), Owen (2006), Peigné et al. (2006) and Labs Hochstein (2007).

Family tree

The cladogram of Arctoidea is as follows:

Source (edited): "http://en.wikipedia.org/wiki/Arctoidea"

Arctonasua

Arctonasua is an extinct genus of raccoon-like procyonid of the Miocene, endemic to North America living from ~17.3—8.4 Mya, existing for approximately 8.9 million years.

Mustelidae include the raccoons, coatis, kinkajous, olingos, ringtails and cacomistles. Procyonids inhabit a wide range of environments, and are generally omnivorous.

Species

A. eurybates

A. eurybates was named by Baskin (1982). Its type locality is ZXBar, which is in a Hemphillian terrestrial horizon in the Snake Creek Formation of Nebraska.

Body mass

A single specimen was examined by Legendre and Roth for body mass.

- Specimen: 17.3 kg (38 lb).

Fossil distribution

- ZXBar site aka Cook's Pliohippus Quarry, Sioux County, Nebraska, estimated age: ~10.5—10.4 Ma.

A. floridana

A. floridana was named by Baskin (1982) as a genotype. Its type locality is Love Bone Bed, which is in a Clarendonian marginal marine horizon in the Alachua Formation of Florida. It is the type species of Arctonasua.

Body mass

A single specimen was examined by Legendre and Roth for body mass.

- Specimen: 13.2 kg (29 lb)

Fossil distribution

- Love Bone Bed site, Alachua County, Florida, estimated age: ~11.5 Ma.

A. fricki

A. fricki was named by Baskin (1982).

Fossil distribution

- Optima site aka Guymon, OMNH V5, or UCMP V-2902, Texas County, Oklahoma, estimated age: ~8.4 Ma.

A. gracilis

A. gracilis was named by Baskin (1982).

Body mass

A single specimen was examined by Legendre and Roth for body mass.

- Specimen: 6.4 kg (14 lb).

Fossil distribution

- Egelhoff Quarry aka UNSM Kp-101, Brown County, Nebraska estimated age: ~8.4 Ma.
- Trinity River Pit 1 (TMM 40196), San Jacinto County, Texas,

estimated age: ~17.3 Ma.

A. minima

A minima was named by Baskin (1982).

Body mass

A single specimen was examined by Legendre and Roth for body mass.

- Specimen: 5.28 kg (12 lb).

Fossil distribution

- Observation Quarry, Dawes County, Nebraska, estimated age: ~17.1 Ma.

Source (edited): "http://en.wikipedia.org/wiki/Arctonasua"

Argyrolagus

Argyrolagus is an extinct genus of South American marsupial.

Jumping on its hind legs, the 40 centimetres (16 in) long *Argyrolagus* resembled a gerbil or kultarr. It had a long tail for balance, and a narrow head with a pointed snout. Judging from its huge eyes, *Argyrolagus* was nocturnal. The form of its teeth suggest that it would have fed on desert plants.

Source (edited): "http://en.wikipedia.org/wiki/Argyrolagus"

Astrapotherium magnum

Astrapotherium magnum ("Great Lightning Beast") is an extinct South American mammal which vaguely resembled a cross between a small elephant, and a very large tapir. It may have fed on marsh plants. This peculiar-looking animal was unrelated to modern elephants, and was, instead, related to other extinct South American ungulates.

Restoration of the head

Astrapotherium had an elongated body, with a total length of about 2.5 metres (8.2 ft) and relatively short limbs. It had small, plantigrade, feet, and the hind limbs were significantly weaker than the fore limbs. Its four canine teeth were elongated to form short tusks, and it had broad protruding lower incisors, which likely ground against a horny pad in the upper jaw, as in many modern ruminants.

The nostrils were placed high on the head, which might indicate the presence of a trunk, but could equally be due to other reasons, such as an inflatable nasal cavity. The animal was probably at least partially aquatic, living in shallow water and feeding on plants in a similar manner to a modern hippopotamus.

Cultural depictions

Mentioned in the book *Ghost* by Piers Anthony.

Source (edited): "http://en.wikipedia.org/wiki/Astrapotherium_magnum"

Aulophyseter

Aulophyseter is an extinct genus of sperm whale from the Miocene formations of the West and East coasts of North America as well as the Patagonian region of South America. It is a possible synonym of some fossils assigned to ***Orycterocetus***.

Related species

- *Orycterocetus*
- *Acrophyseter*
- *Zygophyseter*
- *Brygmophyseter*
- *Livyatan melvillei*

Source (edited): "http://en.wikipedia.org/wiki/Aulophyseter"

Australocamelus

Australocamelus is an extinct genus of terrestrial herbivore the family Camelidae, endemic to North America during the Miocene 16.3—13.6 mya existing for approximately 2.7 million years.

Taxonomy

Australocamelus was named by Patton (1969). Its type is Australocamelus orarius. It was synonymized subjectively with Oxydactylus by Tedford et al. (1987). It was assigned to Camelidae by Patton (1969), Tedford and Barghoorn (1993) and Honey et al. (1998).

Morphology

A single specimen was examined for estimated body mass by M. Mendoza, C. M. Janis, and P. Palmqvist. The specimen was estimated to weigh 146.8 kg (320 lb).

Source (edited): "http://en.wikipedia.org/wiki/Australocamelus"

Barbourofelidae

Barbourofelidae is an extinct family of mammalian carnivores of the suborder Feliformia, which lived in North America, Eurasia and Africa during the Miocene epoch (16.9—9.0 mya), existing for approximately 7.9 million years.

Taxonomy

Barbourofelinae was named by Schultz et al. (1970). It is not extant. Its type is *Barbourofelis*. It was re-ranked as the family **Barbourofelidae** by Morlo et al. (2004). It was assigned to Nimravidae by Bryant (1991); and to Carnivora by Morlo et al. (2004).

Barbourofelidae were previously placed as a subfamily of the extinct Nimravidae, the barbourofelids have been recently assigned to their own distinct family, and are now thought to be taxonomically closer to the Felidae than to the Nimravidae. They first appear in the fossil record in the early Miocene of Africa. By the end of the early Miocene a land bridge had opened between Africa and Eurasia, allowing for a faunal exchange between the two continents. Barbourofelids migrated at least three times from Africa to Europe (Morlo 2006) . While evolving in Europe to the genus *Sansanosmilus*, Barbourofelids migrated also through Eurasia and reached North America by the late Miocene, represented there by the genus *Barbourofelis* only.

Classification

- Family †**Barbourofelidae**
  - Genus †*Ginsburgsmilus*
    - †*Ginsburgsmilus napakensis*
  - Genus †*Afrosmilus*
    - †*Afrosmilus turkanae*
    - †*Afrosmilus africanus*
    - †*Afrosmilus hispanicus*
  - Genus †*Prosansanosmilus*
    - †*Prosansanosmilus peregrinus*
    - †*Prosansanosmilus eggeri*
  - Genus †*Sansanosmilus*
    - †*Sansanosmilus palmidens*
    - †*Sansanosmilus jourdani*
    - †*Sansanosmilus vallesiensis*
    - †*Sansanosmilus piveteaui*
  - Genus †*Syrtosmilus*
    - †*Syrtosmilus syrtensis*
  - Genus †*Vampyrictis*
    - †*Vampyrictis vipera*
  - Genus †*Barbourofelis*
    - †*Barbourofelis whitfordi*
    - †*Barbourofelis lovei*
    - †*Barbourofelis morrisi*
    - †*Barbourofelis fricki*

Source (edited): "http://en.wikipedia.org/wiki/Barbourofelidae"

Barbourofelis

Barbourofelis is an extinct genus of large, mostly carnivorous mammals of the family Barbourofelidae (false sabertooth cats) endemic to North America, during the Miocene living from 13.6—5.3 Ma and existed for approximately 8.3 million years.

Taxonomy

B. loveorum skeleton mounted as if attacking *Teleoceras proterum*

Barbourofelis was named by Schultz and et al. (1970). Its type is Barbourofelis fricki. It is the type genus of Barbourofelinae. It was assigned to Hoplophoneinae by Flynn and Galiano (1982); to Barbourofelinae by Bryant (1991); and to Nimravidae by Schultz and et al. (1970) and Martin (1998).

Morphology

A single specimen was examined by Legendre and Roth for body mass. The first specimen was estimated to weigh 66.4 kg (150 lb).

Thought to possibly be lion-sized and had the longest canines of all barbourofelids. They had very prominent flanges on the lower jaws and an unusually shaped skull. The barbourofelids were probably very muscular, resembling a bear-like lion or lion-like bear. Although the nimravid family did evolve into cat-like forms, they left no descendents among modern cats.

Source (edited): "http://en.wikipedia. org/wiki/Barbourofelis"

Barbourofelis

Barbourofelis is an extinct genus of large, mostly carnivorous mammals of the family Barbourofelidae (false saber-tooth cats) endemic to North America, during the Miocene living from 13.6—5.3 Ma and existed for approximately 8.3 million years.

Taxonomy

B. loveorum skeleton mounted as if attacking *Teleoceras proterum*

Barbourofelis was named by Schultz and et al. (1970). Its type is Barbourofelis fricki. It is the type genus of Barbourofelinae. It was assigned to Hoplophoneinae by Flynn and Galiano (1982); to Barbourofelinae by Bryant (1991); and to Nimravidae by Schultz and et al. (1970) and Martin (1998).

Morphology

A single specimen was examined by Legendre and Roth for body mass. The first specimen was estimated to weigh 66.4 kg (150 lb).

Thought to possibly be lion-sized and had the longest canines of all barbourofelids. They had very prominent flanges on the lower jaws and an unusually shaped skull. The barbourofelids were probably very muscular, resembling a bear-like lion or lion-like bear. Although the nimravid family did evolve into cat-like forms, they left no descendents among modern cats.

Source (edited): "http://en.wikipedia. org/wiki/Barbourofelis"

Barinya

Barinya is a fossil genus from the marsupial family Dasyuridae, which contains the oldest known undoubted dasyurid.

The principal difference between *Barinya* and more recent dasyurids are in the dentition and skull morphology; with *Barinya* displaying more primitive features. There is one described fossil and at least one that remains to be described. This genus has only been found at Riversleigh in Queensland where it is quite common in deposits from the Oligo-Miocene.

Source (edited): "http://en.wikipedia. org/wiki/Barinya"

Bathyergoides

Bathyergoides is an extinct genus of rodent from Africa thought to be related to the modern blesmols. It is the only member of the family **Bathyergoididae**.

Source (edited): "http://en.wikipedia. org/wiki/Bathyergoides"

Bear dog

Amphicyonidae is an extinct family of large terrestrial carnivores belonging to the suborder Caniformia (meaning "dog-like") and which inhabited North America, Europe, Asia, and Africa from the Middle Eocene subepoch to the Pleistocene epoch 46.2—1.8 Mya, existing for approximately 44.4 million years.

Taxonomy

An artist's reconstruction of a bear dog

Amphicyonidae was named by Haeckel (1886) [also attributed to Trouessart 1885]. It was assigned to Carnivora by Sach and Heizmann (2001); to Arctoidea by Hunt (2001), Hunt (2002) and Hunt (2002); affirmed as Arctoidea by Zhai et al. (2003); affirmed to Arctoidea by Carroll (1988), Hunt (1998) and Wang et al. (2005); and to Caniformia by Morlo et al. (2007).

Origins

Amphicyonids, often referred to as "bear dogs", crossed from Europe to North America during the Miocene epoch and are considered an Old World taxon. The earliest to appear is the (rather large) *Ysengrinia* (30—20 Mya), followed by *Cynelos* (24—7 Mya) and *Amphicyon* (23—5 Mya). These animals would have followed ungulates and other mammals to the New World for a period of approximately 7 million years. The New World amphicyonids of the subfamilies Daphoeninae (42-16 Mya) and Temnocyoninae (33-20 Mya) coexisted with the Old World counterparts. Note that the (often similar looking) members of the family Hemicyonidae are often called "bear-dogs" as well (although they are increasingly referred to as "dog-bears" to avoid confusion).

Amphicyonids were as small as 5 kilograms (11 lb) and as large as 100 to 600 kilograms (220 to 1,300 lb) and evolved from wolf-like to bear-like. The diet of the amphicyonids was fully carnivorous as opposed to hypercarnivorous to mesocarnivorous in Canidae.

Evolution

While amphicyonids have traditionally been viewed as closely related to ursids (bears), some evidence suggests that they may instead be basal caniforms. (Hunt, 2004b). They were about as tall as the American black bear and were most likely ambushers because their legs were made for short, sudden bursts of speed. Bear-dog also nested their young in underground burrows.

During the early Miocene, a number of large amphicyonids migrated from Eurasia into North America. These taxa belong to the Old World amphicyonid sub-family Amphicyoninae. The earliest to appear is the large bear dog *Ysengrinia* Ginsburg, followed by *Cynelos* Jourdan, and then by *Amphicyon*. This influx of amphicyonines, accompanied by other Old World ungulates and small mammals, indicates a prolonged interval (from 23 to 16.5 Ma) of faunal exchange between Asia and North America in the early Miocene, using the trans-Beringian route. New World daphoenines (*Daphoenodon*, *Borocyon*) and temnocyonines coexisted with Old World amphicyonines (*Ysengrinia*, *Amphicyon*, *Cynelos*) 23.7-17.5 million years ago. These are the largest terrestrial carnivorans 50 kilograms (110 lb) to 200 kilograms (440 lb) that evolved on the North American continent up to this time. The immigrant amphicyonines *Ysengrinia*, *Cynelos* and *Amphicyon* appear at 23, 19.2, and 18.8 Ma, respectively, and herald the beginning of a Eurasian amphicyonine migration into North America that continued into the mid-Miocene.

Classification

- **Family Amphicyonidae**
  - **Subfamily Amphicyoninae**
    - **Genus *Agnotherium***
      - *A. antiquus*
      - *A. grivense*
    - **Genus *Amphicyon***
      - *A. frendens*
      - *A. bohemicus*
      - *A. castellanus*
      - *A. caucasicus*
      - *A. galushai*
      - *A. giganteus*
      - *A. ingens*
      - *A. intermedius (type)*
      - *A. laugnacensis*
      - *A. longiramus*
      - *A. major*
      - *A. pontoni*
      - *A. reinheimeri*
      - *A. riggsi*
      - *A. tairumensis*
      - *A. ulungurensis*
    - **Genus *Amphicyonopsis***
      - *A. serus*
    - **Genus *Brachycyon***
      - *B. reyi*
      - *B. palaeolycos*
      - *B. gaudryi*
    - **Genus *Cynelos***
      - *C. caroniavorus*
      - *C. crassidens*
      - *C. helbingo*
      - *C. idoneus*
      - *C. jourdan*
      - *C. lemanensis*
      - *C. pivetaui*
      - *C. rugosidens*
      - *C. schlosseri*
      - *C. sinapius*
    - **Genus *Cynodictis***
      - *C. lacustris*
    - **Genus *Euroamphicyon***
      - *E. olisiponensis*
    - **Genus *Gobicyon***
      - *G. macrognathus*
      - *G. zhegalloi*
    - **Genus *Guangxicyon***
      - *G. sinoamericanus*
    - **Genus *Haplocyon***
      - *H. elegans*
      - *H. crucians*
    - **Genus *Haplocyonoides***
      - *H. mordax*
      - *H. serbiae*
      - *H. ponticus*
    - **Genus *Haplocyonopsis***
    - **Genus *Harpagocyon***
    - **Genus *Heducides***
    - **Genus *Ischyrocyon***
      - *I. gidleyi*
    - **Genus *Paradaphoenus***
      - *P. cuspigerus*
      - *P. minimus*
      - *P. tooheyi*
    - **Genus *Pericyon***
    - **Genus *Pliocyon***
      - *P. medius*
      - *P. robustus*
    - **Genus *Proamphicyon***
    - **Genus *Protemnocyon***
    - **Genus *Pseudarctos***
      - *P. bavaricus*
    - **Genus *Pseudamphicyon***
      - *P. bavaricus*
    - **Genus *Pseudocyon***
      - *P. sansaniensis*
      - *P. steinheimensis*
      - *P. styriacus*
    - **Genus *Pseudocyonopsis***
      - *P. ambiguus*
      - *P. antiquus*
      - *P. quercensis*
    - **Genus *Symplectocyon***
    - **Genus *Ysengrinia***
      - *Y. americanus*
      - *Y. depereti*
      - *Y. geraniana*
      - *Y. ginsburg*
      - *Y. tolosana*
  - **Subfamily Daphoeninae**(North America)
    - **Genus *Adilophontes***

- *A. brachykolos*
- **Genus *Borocyon***
- **Genus *Brachyrhyncocyon***
  - *B. dodgei*
  - *B. montanus*
- **Genus *Daphoenictis***
  - *D. tedfordi*
- **Genus *Daphoenodon***
  - *D. falkenbachi*
  - *D. notionastes*
  - *D. robustum*
  - *D. periculosus*
  - *D. skinneri*
  - *D. superbus*
- **Genus *Daphoenus***
  - *D. dodgei* or *Daphoenocyon dodgei*
  - *D. felinus*
  - *D. hartshornianus*
  - *D. inflatus*
  - *D. lambei*
  - *D. nebrascensis*
  - *D. socialis*
  - *D. transversus*
  - *D. vetus*
- **Genus *Paradaphoenus***
  - *P. cuspigerus*
  - *P. minimus*
  - *P. tooheyi*
- **Subfamily Temnocyoninae** (North America)
  - **Genus *Mammacyon***
    - *M. obtusidens*
  - **Genus *Temnocyon***
    - *T. altigenis*
    - *T. ferox*
    - *T. percussor*
    - *T. venator*
- **Subfamily Thanmastocyoninae**

Source (edited): "http://en.wikipedia.org/wiki/Bear_dog"

Blastomeryx

Blastomeryx is an extinct genus of musk deer of the family *Moschidae*, subfamily *Blastomerycinae*, endemic to North America during the Oligocene-Miocene epochs (24.8—10.3 mya), existing for approximately 14.5 million years.

Morphology

Restoration

It was 75 cm (2 ft 6 in) long and looked like a modern chevrotain. Its canines were elongated into tusks which it probably used to uproot plants and fend off predators. While most species of *Blastomeryx* (as well as modern musk deer) lacked antlers, a Late Miocene species had bony knobs on its skull, which have been interpreted as incipient horns.

Body mass

Two specimens was examined by M. Mendoza for body mass. The first specimen was estimated to have a weight of 55.3 kg (120 lb). The second specimen was estimated to have a weight of 16.1 kg (35 lb).

Source (edited): "http://en.wikipedia.org/wiki/Blastomeryx"

Borhyaena

Borhyaena is an extinct genus of South American marsupial, living between 20 and 15 million years ago.

Borhyaena was a large, bear-like marsupial predator of up to 1.5 metres (5 ft) long and with an estimated average weight of 100 kilograms (220 lb). A typical borhyaenid, it had flat feet and a heavy build. *Borhyaena* probably hunted from an ambush as its short legs would not have allowed it to run very fast.

Source (edited): "http://en.wikipedia.org/wiki/Borhyaena"

Borophagus diversidens

Borophagus diversidens ("devouring glutton") is an extinct species of the genus *Borophagus* of the subfamily *Borophaginae,* a group of canids endemic to North America from the early Miocene epoch through the Pliocene epoch 4.9—1.8 Ma. *Borophagus dudleyi* existed for approximately 3.1 million years.

Overview

Borophagus diversidens was named by Cope in 1892. *Borophagus diversidens*, like other *Borophaginae*, are loosely known as "bone-crushing" or "hyena-like" dogs. Though not the most massive borophagine by size or weight, it had a more highly-evolved capacity to crunch bone than earlier, larger genera such as *Epicyon*, which seems to be an evolutionary trend of the group (Turner, 2004). During the Pliocene epoch, *Borophagus* began being displaced by *Canis* genera such as *Canis edwardii* and later by *Canis dirus*. Early species of *Borophagus* were placed in the genus ***Osteoborus*** until recently, but the genera are now considered synonyms. *Borophagus diversidens* possibly led a hyena-like lifestyle scavenging carcasses of recently dead animals.

Taxonomy

Typical features of this genus are a bulging forehead and powerful jaws; it was probably a scavenger. Its crushing premolar teeth and strong jaw muscles would have been used to crack open bone, much like the hyena of the Old World. The adult animal is estimated to have been about 80 cm in length, similar to a coyote, although it was much more powerfully built.

Morphology

Two fossil specimens of *Borophagus diversidens* were measured by Legendre and Roth in 1988. They estimated that specimen one weighed 89 kg (196 lbs) and the second weighed 75.4 kg (166 lbs).

Recombination

Borophagus diversidens was recombined as *Dinocyon Borophagus diversidens* by Matthew in 1902 and then recombined as *Dinocyon diversidens* by Matthew the same year. It was recombined as *Hyaenarctos diversidens*. It has been synonymized with *Felis hillianus*, *Hyaenognathus matthewi*, *Hyaenognathus pachyodon*, *Hyaenognathus solus*, and *Porthocyon dubius*.

Fossil distribution

Borophagus diversidens fossil specimens are very widespread from 2 sites in central Florida to central Mexico, from western Oregon and western Washington to New Mexico, Arizona, and Texas.

Species

- *Borophagus dudleyi* existed for 19.7 million years
- *Borophagus hilli* existed for 19.7 million years (synonymous with *Osteoborus crassapineatus*, *Osteoborus progressus*)
- *Borophagus littoralis* existed for 13 million years (syn. Osteoborus diabloensis)
- *Borophagus orc* existed for 5.4 million years
- *Borophagus parvus* existed for 6.4 million years
- *Borophagus pugnator* existed for 18 million years (synonymous with *Osteoborus galushai*)
- *Borophagus secundus* existed for 18 million years (synonymous with *Hyaenognathus cyonoides*, *Hyaenognathus direptor*)

Existence based on age of fossil collections and recombination with other species.

Sister genera

Carpocyon, *Epicyon*, *Paratomarctus*, *Protepicyon*

Source (edited): "http://en.wikipedia.org/wiki/Borophagus_diversidens"

Borophagus dudleyi

Borophagus dudleyi ("devouring glutton") is an extinct species of the genus *Borophagus* of the subfamily *Borophaginae,* a group of canids endemic to North America from the late Hemphillian of the Miocene epoch (10.3 Mya) through the Pliocene epoch (4.9 Mya). *Borophagus dudleyi* existed for approximately 19.7 million years.

Overview

Borophagus dudleyi was originally named *Pliogulo dudleyi* by T.E. White in 1941. *Borophagus dudleyi*, like other *Borophaginae*, are loosely known as "bone-crushing" or "hyena-like" dogs. Though not the most massive borophagine by size or weight, it had a more highly-evolved capacity to crunch bone than earlier, larger genera such as *Epicyon*, which seems to be an evolutionary trend of the group (Turner, 2004). During the Pliocene epoch, *Borophagus* began being displaced by *Canis* genera such as *Canis edwardii* and later by *Canis dirus*. Early species of *Borophagus* were placed in the genus ***Osteoborus*** until recently, but the genera are now considered synonyms. *Borophagus parvus* possibly led a hyena-like lifestyle scavenging carcasses of recently dead animals.

Taxonomy

Borophagus dudleyi was recombined as *Cynogulo dudleyi* by Kretzoi in 1968. It was recombined again as *Osteoborus dudleyi* by Webb in 1969 and Munthe in 1998. The animal was then synonymized subjectively with *Borophagus crassapineatus* by Richey in 1979. In 1987, Richard H. Tedford returned its original name and Xiaoming Wang along with Richard H. Tedford and Beryl E. Taylor concurred in a 1999 examination.

Morphology

Typical features of this genus are a bulging forehead and powerful jaws; it was probably a scavenger. Its crushing premolar teeth and strong jaw muscles would have been used to crack open bone, much like the hyena of the Old World. The adult animal is estimated to have been about 80 cm in length, similar to a coyote, although it was much more powerfully built.

Fossil distribution

Borophagus dudleyi fossil specimens are exclusive to a coastal area of North Carolina.

Species

- *Borophagus diversidens* existed for 3.1 million years (synonymous with *Felis hillianus*, *Hyaenognathus matthewi*, *Hyaenognathus pachyodon*, *Hyaenognathus solus*, *Porthocyon dubius*)
- *Borophagus hilli* existed for 19.7 million years (synonymous with *Osteoborus crassapineatus*, *Osteoborus progressus*)
- *Borophagus littoralis* existed for 13 million years (syn. Osteoborus diabloensis)
- *Borophagus orc* existed for 5.4 million years
- *Borophagus parvus* existed for 6.4

million years
- *Borophagus pugnator* existed for 18 million years (synonymous with *Osteoborus galushai*)
- *Borophagus secundus* existed for 18 million years (synonymous with *Hyaenognathus cyonoides*, *Hyaenognathus direptor*)

Existence based on age of fossil collections and recombination with other species.

Sister genera

Carpocyon, *Epicyon*, *Paratomarctus*, *Protepicyon*

Source (edited): "http://en.wikipedia.org/wiki/Borophagus_dudleyi"

Borophagus hilli

Borophagus hilli ("devouring glutton") is an extinct species of the genus *Borophagus* of the subfamily *Borophaginae,* a group of canids endemic to North America from the early Miocene epoch (23.3 Mya) through the Pliocene epoch (4.9 Mya). *Borophagus dudleyi* existed for approximately 19.7 million years.

Overview

Borophagus hilli was named by C. S. Johnston in 1939. *Borophagus dudleyi*, like other *Borophaginae*, are loosely known as "bone-crushing" or "hyena-like" dogs. Though not the most massive borophagine by size or weight, it had a more highly-evolved capacity to crunch bone than earlier, larger genera such as *Epicyon*, which seems to be an evolutionary trend of the group (Turner, 2004). During the Pliocene epoch, *Borophagus* began being displaced by *Canis* genera such as *Canis edwardii* and later by *Canis dirus*. Early species of *Borophagus* were placed in the genus ***Osteoborus*** until recently, but the genera are now considered synonyms. *Borophagus hilli* possibly led a hyena-like lifestyle scavenging carcasses of recently dead animals.

Taxonomy

Typical features of this genus are a bulging forehead and powerful jaws; it was probably a scavenger. Its crushing premolar teeth and strong jaw muscles would have been used to crack open bone, much like the hyena of the Old World. The adult animal is estimated to have been about 80 cm in length, similar to a coyote, although it was much more powerfully built.

Morphology

Two fossil specimens of *Borophagus hilli* were measured by Legendre and Roth for body mass.
- Specimen 1: 61 kg (130 lb)
- Specimen 2: 57.7 kg (130 lb)

Recombination

Borophagus hilli was synonymized subjectively with *Borophagus direptor* by Kurten and Anderson in 1980 as well as synonymous with *Osteoborus crassapineatus*, *Osteoborus progressus*. It was recombined as *Borophagus hilli* by Xiaoming Wang et al. in 1999.

Fossil distribution

Borophagus hilli fossil specimens are widespread from east central Florida to southeastern Washington, from Idaho to New Mexico to Texas. Specimens were also found as far south as the southern tip of Baja, Mexico

Species

- *Borophagus dudleyi* existed for 19.7 million years
- *Borophagus diversidens* existed for 3.1 million years (synonymous with *Felis hillianus*, *Hyaenognathus matthewi*, *Hyaenognathus pachyodon*, *Hyaenognathus solus*, *Porthocyon dubius*)
- *Borophagus littoralis* existed for 13 million years (syn. Osteoborus diabloensis)
- *Borophagus orc* existed for 5.4 million years
- *Borophagus parvus* existed for 6.4 million years
- *Borophagus pugnator* existed for 18 million years (synonymous with *Osteoborus galushai*)
- *Borophagus secundus* existed for 18 million years (synonymous with *Hyaenognathus cyonoides*, *Hyaenognathus direptor*)

Existence based on age of fossil collections and recombination with other species.

Sister genera

Carpocyon, *Epicyon*, *Paratomarctus*, *Protepicyon*

Source (edited): "http://en.wikipedia.org/wiki/Dorophagus_hilli"

Borophagus littoralis

Borophagus littoralis ("devouring glutton") is an extinct species of the genus *Borophagus* of the subfamily *Borophaginae,* a group of canids endemic to North America from the early Miocene epoch (23.3 Mya) through the Pliocene epoch (4.9 Mya). *Borophagus littoralis* existed for approximately 10 million years.

Overview

Borophagus littoralis was named by Cope in 1892 and is considered synonomized with Osteoborus diabloensis. *Borophagus littoralis, like other* Borophaginae, *are loosely known as "bone-crushing" or "hyena-like" dogs. Though not the most massive borophagine by size or weight, it had a more highly-evolved capacity to crunch bone than earlier, larger genera such as* Epicyon, *which seems to be an evolutionary trend of the group (Turner, 2004). During the Pliocene epoch,*

Borophagus *began being displaced by* Canis *genera such as* Canis edwardii *and later by* Canis dirus. *Early species of* Borophagus *were placed in the genus* **Osteoborus** *until recently, but the genera are now considered synonyms.* Borophagus littoralis *possibly led a hyena-like lifestyle scavenging carcasses of recently dead animals.*

Taxonomy

Typical features of this genus are a bulging forehead and powerful jaws; it was probably a scavenger. Its crushing premolar teeth and strong jaw muscles would have been used to crack open bone, much like the hyena of the Old World. The adult animal is estimated to have been about 80 cm in length, similar to a coyote, although it was much more powerfully built.

Morphology

Two fossil specimens of *Borophagus littoralis* were measured by Legendre and Roth for body mass.

- Specimen 1: 40.2 kg (89 lb)
- Specimen 2: 35.4 kg (78 lb)

Fossil distribution

Borophagus diversidens fossil specimens are confined to 3 sites in California.

Species

- *Borophagus dudleyi* existed for 19.7 million years
- *Borophagus hilli* existed for 19.7 million years (synonymous with *Osteoborus crassapineatus*, *Osteoborus progressus*)
- *Borophagus orc* existed for 5.4 million years
- *Borophagus parvus* existed for 6.4 million years
- *Borophagus pugnator* existed for 18 million years (synonymous with *Osteoborus galushai*)
- *Borophagus secundus* existed for 18 million years (synonymous with *Hyaenognathus cyonoides*, *Hyaenognathus direptor*)

Existence based on age of fossil collections and recombination with other species.

Sister genera

Carpocyon, *Epicyon*, *Paratomarctus*, *Protepicyon*

Source (edited): "http://en.wikipedia.org/wiki/Borophagus_littoralis"

Borophagus orc

Borophagus orc is an extinct species of the genus *Borophagus* of the subfamily *Borophaginae,* a group of canids endemic to North America from the late Zanclean stage of the Pliocene epoch (10.3 Mya) through the Hemphillian stage of the Miocene epoch (4.9 Mya). *Borophagus orc* existed for approximately 5.4 million years.

Overview

Borophagus, like other *Borophaginae*, are loosely known as "bone-crushing" or "hyena-like" dogs. Though not the most massive borophagine by size or weight, it had a more highly-evolved capacity to crunch bone than earlier, larger genera such as *Epicyon*, which seems to be an evolutionary trend of the group (Turner, 2004). During the Pliocene epoch, *Borophagus* began being displaced by *Canis* genera such as *Canis edwardii* and later by *Canis dirus*. Early species of *Borophagus* were placed in the genus ***Osteoborus*** until recently, but the genera are now considered synonyms. *Borophagus orc* possibly led a hyena-like lifestyle scavenging carcasses of recently dead animals.

Taxonomy

Typical features of this genus are a bulging forehead and powerful jaws; it was probably a scavenger. Its crushing premolar teeth and strong jaw muscles would have been used to crack open bone, much like the hyena of the Old World. The adult animal is estimated to have been about 80 cm in length, similar to a coyote, although it was much more powerfully built.

Morphology

Two fossil specimens of *Borophagus orc* were measured by Legendre and Roth in 1988. They estimated that specimen one weighed 24.3 kg (53.5 lbs) and the second weighed 22 kg (48.5 lbs).

Recombination

Borophagus orc was recombined by X. Wang in 1999.

Fossil distribution

Specimens were found at only 2 sites. Near Withlacoochee River, Florida and coastal North Carolina.

Species

- *Borophagus diversidens* existed for 3.1 million years (synonymous with *Felis hillianus*, *Hyaenognathus matthewi*, *Hyaenognathus pachyodon*, *Hyaenognathus solus*, *Porthocyon dubius*)
- *Borophagus dudleyi* existed for 19.7 million years
- *Borophagus hilli* existed for 19.7 million years (synonymous with *Osteoborus crassapineatus*, *Osteoborus progressus*)
- *Borophagus littoralis* existed for 13 million years (syn. Osteoborus diabloensis)
- *Borophagus parvus* existed for 6.4 million years
- *Borophagus pugnator* existed for 18 million years (synonymous with *Osteoborus galushai*)
- *Borophagus secundus* existed for 18 million years (synonymous with *Hyaenognathus cyonoides*, *Hyaenognathus direptor*)

Existence based on age of fossil collections and recombination with other species.

Sister genera

Carpocyon, Epicyon, Paratomarctus, Protepicyon

Source (edited): "http://en.wikipedia.org/wiki/Borophagus_orc"

Borophagus parvus

Borophagus parvus ("devouring glutton") is an extinct species of the genus *Borophagus,* of the subfamily *Borophaginae,* a group of canids (Bear-dog) endemic to North America from the late Hemphillian of the Miocene epoch through the Pliocene epoch 10.3—4.9 Ma., existing for approximately 5.4 million years.

Overview

Borophagus, like other *Borophaginae*, are loosely known as "bone-crushing" or "hyena-like" dogs. Though not the most massive borophagine by size or weight, it had a more highly-evolved capacity to crunch bone than earlier, larger genera such as *Epicyon*, which seems to be an evolutionary trend of the group (Turner, 2004). During the Pliocene epoch, *Borophagus* began being displaced by *Canis* genera such as *Canis edwardii* and later by *Canis dirus*. Early species of *Borophagus* were placed in the genus ***Osteoborus*** until recently, but the genera are now considered synonyms. *Borophagus parvus* possibly led a hyena-like lifestyle scavenging carcasses of recently dead animals.

Taxonomy

Paracynarctus was named by Wang et al. (1999). Its type is Paracynarctus sinclairi. It was assigned to Cynarctina by Wang et al. (1999).

Morphology

Typical features of this genus are a bulging forehead and powerful jaws; it was probably a scavenger. Its crushing premolar teeth and strong jaw muscles would have been used to crack open bone, much like the hyena of the Old World. The adult animal is estimated to have been about 80 cm in length, similar to a coyote, although it was much more powerfully built.

Two fossil specimens of *Borophagus parvus* were measured by Legendre and Roth in 1988. They estimated that the first specimen weighed 28.8 kg (63 lb) and the second weighed 25.8 kg (57 lb).

Species

- *Borophagus diversidens* existed for 3.1 million years (synonymous with *Felis hillianus, Hyaenognathus matthewi, Hyaenognathus pachyodon, Hyaenognathus solus, Porthocyon dubius*)
- *Borophagus dudleyi* existed for 19.7 million years
- *Borophagus hilli* existed for 19.7 million years (synonymous with *Osteoborus crassapineatus, Osteoborus progressus*)
- *Borophagus littoralis* existed for 13 million years (syn. Osteoborus diabloensis)
- *Borophagus orc* existed for 5.4 million years
- *Borophagus pugnator* existed for 18 million years (synonymous with *Osteoborus galushai*)
- *Borophagus secundus* existed for 18 million years (synonymous with *Hyaenognathus cyonoides, Hyaenognathus direptor*)

Existence based on age of fossil collections and recombination with other species.

Sister genera

Carpocyon, Epicyon, Paratomarctus, Protepicyon

Source (edited): "http://en.wikipedia.org/wiki/Borophagus_parvus"

Borophagus pugnator

Borophagus pugnator ("devouring glutton") is an extinct species of the genus *Borophagus* of the subfamily *Borophaginae,* a group of canids endemic to North America from the early Miocene epoch through the late Miocene epoch 23.3—4.9 Ma. *Borophagus pugnator* existed for approximately 18 million years.

Overview

Borophagus, like other *Borophaginae*, are loosely known as "bone-crushing" or "hyena-like" dogs. Though not the most massive borophagine by size or weight, it had a more highly-evolved capacity to crunch bone than earlier, larger genera such as *Epicyon*, which seems to be an evolutionary trend of the group (Turner, 2004). During the Pliocene epoch, *Borophagus* began being displaced by *Canis* genera such as *Canis edwardii* and later by *Canis dirus*. Early species of *Borophagus* were placed in the genus ***Osteoborus*** until recently, but the genera are now considered synonyms. *Borophagus pugnator* possibly led a hyena-like lifestyle scavenging carcasses of recently dead animals.

Taxonomy

Typical features of this genus are a bulging forehead and powerful jaws; it was probably a scavenger. Its crushing premolar teeth and strong jaw muscles would have been used to crack open bone, much like the hyena of the Old World. The adult animal is estimated to have been about 80 cm in length, similar to a coyote, although it was much more powerfully built.

Morphology

Two fossil specimens of *Borophagus pugnator* were measured by Legendre and Roth for body mass. The results are as follows:

- Specimen 1: 41.1 kg (91 lb)
- Specimen 2: 36 kg (79 lb)

Recombination

Borophagus pugnator was originally named *Porthocyon pugnator* by Cook 1932. It was recombined by Matthew as *Aelurodon pugnator* in 1924 followed by Matthew and Stirton in 1830. It was then recombined as *Osteoborus pugnator* by Stirton and VanderHoof in 1933 and others until recombined as *Borophagus pugnator* by VanderHoof (1931) and X. Wang et al. in 1999.

Fossil distribution

Specimens were found at only 2 sites. Near Withlacoochee River, Florida and coastal North Carolina.

Species

- *Borophagus diversidens* existed for 3.1 million years (synonymous with *Felis hillianus, Hyaenognathus matthewi, Hyaenognathus pachyodon, Hyaenognathus solus, Porthocyon dubius*)
- *Borophagus dudleyi* existed for 19.7 million years
- *Borophagus hilli* existed for 19.7 million years (synonymous with *Osteoborus crassapineatus, Osteoborus progressus*)
- *Borophagus littoralis* existed for 13 million years (syn. Osteoborus diabloensis)
- *Borophagus parvus* existed for 6.4 million years
- *Borophagus secundus* existed for 18 million years (synonymous with *Hyaenognathus cyonoides, Hyaenognathus direptor*)

Existence based on age of fossil collections and recombination with other species.

Sister genera

Carpocyon, Epicyon, Paratomarctus, Protepicyon

Source (edited): "http://en.wikipedia.org/wiki/Borophagus_pugnator"

Borophagus secundus

Borophagus secundus ("devouring glutton") is an extinct species of the genus *Borophagus* of the subfamily *Borophaginae,* a group of canids endemic to North America from the Early Miocene epoch (23.0 Mya) through the Late Miocene epoch (5.3 Mya). *Borophagus secundus* existed for approximately 17.7 million years.

Overview

Borophagus secundus, like other *Borophaginae*, are loosely known as "bone-crushing" or "hyena-like" dogs. Though not the most massive borophagine by size or weight, it had a more highly-evolved capacity to crunch bone than earlier, larger genera such as *Epicyon*, which seems to be an evolutionary trend of the group (Turner, 2004). During the Pliocene epoch, *Borophagus* began being displaced by *Canis* genera such as *Canis edwardii* and later by *Canis dirus*. Early species of *Borophagus* were placed in the genus ***Osteoborus*** until recently, but the genera are now considered synonyms. *Borophagus diversidens* possibly led a hyena-like lifestyle scavenging carcasses of recently dead animals.

Taxonomy

Typical features of this genus are a bulging forehead and powerful jaws; it was probably a scavenger. Its crushing premolar teeth and strong jaw muscles would have been used to crack open bone, much like the hyena of the Old World. The adult animal is estimated to have been about 80 cm in length, similar to a coyote, although it was much more powerfully built.

Morphology

Two fossil specimens of *Borophagus secundus* were measured by Legendre and Roth in 1988. They estimated that specimen one weighed 41 kg (90.3 lbs) and the second weighed 36.1 kg (79.5 lbs).

Fossil distribution

Borophagus secundus fossil specimens are very widespread from Honduras and El Salvador to central Mexico, Oklahoma panhandle, central and southern California, Nebraska, Kansas, and northern New Mexico.

Species

- *Borophagus diversidens* existed for 3.1 million years (synonymous with *Felis hillianus, Hyaenognathus matthewi, Hyaenognathus pachyodon, Hyaenognathus solus, Porthocyon dubius*)
- *Borophagus dudleyi* existed for 19.7 million years
- *Borophagus hilli* existed for 19.7 million years (synonymous with *Osteoborus crassapineatus, Osteoborus progressus*)
- *Borophagus littoralis* existed for 13 million years (syn. Osteoborus diabloensis)
- *Borophagus orc* existed for 5.4 million years
- *Borophagus parvus* existed for 6.4 million years
- *Borophagus pugnator* existed for 18 million years (synonymous with *Osteoborus galushai*)

Existence based on age of fossil collections and recombination with other species.

Sister genera

Carpocyon, Epicyon, Paratomarctus, Protepicyon

Source (edited): "http://en.wikipedia.org/wiki/Borophagus_secundus"

Brachycrus

Brachycrus is an extinct genus of terrestrial herbivore of the family *Merycoidodontidae,* subfamily *Merycochoerinae,* (oreodont) endemic to North America during the Middle Miocene to Late Miocene subepochs (20.6—13.6 mya) existing for approximately 7 million years.

Taxonomy

Life restoration of *Brachycrus laticeps*

The 1 metre (3.3 ft) long creature resembled its bigger, earlier relative *Merycoidodon*, but was more specialized. *Brachycrus* had jaws which were short, and because the nostrils were placed far to the back the creature is presumed to have a tapir-like proboscis.

Brachycrus was named by Matthew 1901 as a subgenus of *Merycochoerus* and raised to genus level by Palmer in 1904. It was synonymized subjectively with *Ticholeptus* by Thorpe (1937). It was assigned to *Merycoidodontidae* by Palmer (1904), Schultz and Falkenbach (1940), Schultz and Falkenbach (1968), Kelly and Lander (1988) and Lander (1998).

Morphology

A single specimen was examined for body mass by M. Mendoza and estimated to weigh 328.7 kilograms (725 lb).

Source (edited): "http://en.wikipedia.org/wiki/Brachycrus"

Brygmophyseter

Brygmophyseter

Brygmophyseter is an extinct genus of toothed whale in the sperm whale family with one species ***Brygmophyseter shigensis***. When first described, this species was placed in an extinct form genus *Scadicetus* of toothed whale, as ***Scadicetus shigensis***). In 2006, it was transferred to a new genus *Naganocetus* as "***Naganocetus shigensis***", however later in 2006, it was then moved to *Brygmophyseter* leaving *Naganocetus* as a nomen nudum name.

Etymology

The genus name is a combination of the Greek word *brygmos*, which means "biting" or "gnashing" and "*Physeter*" the living sperm whale genus, and derived from the Greek word meaning "blower". This has led to the occasional vernacular term "biting sperm whale" being given to the species.

Discovery and fossil distribution

The remains of this ancient sperm whale were discovered in Japan by two paleontologists, K. Hirota and L. G. Barnes in 1992, and included a 15 million year old 140 cm long skull. Unlike the modern sperm whale *Physeter catodon* which has teeth in its lower jaw only, *Brygmophyseter* has 12 functional teeth on each side of the lower jaws and 12 functional teeth on each side of the upper jaws.

Size estimation

A nearly complete fossil specimen of this ancient whale from Japan is about 7 meters (23 feet) long. However, paleontologists estimate that *Brygmophyseter* could have grown up to 12 meters (40 feet) long, and dwarfed the modern age Killer Whales.

Behavior and predatory capabilities

Brygmophyseter was likely among the apex predators of its time. It packed a formidable jaws armed with teeth up to 14 cm (5.5 inches) long. It may have employed sonar to track and disorient a prey item in deep waters, like the modern age Sperm Whales. It may also have used its large head to ram a potential opponent during conflict. *Brygmophy seter* would also have lived in social groups called *pods*, and probably hunted in packs, like modern age Orcas.

Source (edited): "http://en.wikipedia.org/wiki/Brygmophyseter"

Bugtirhinus

Bugtirhinus was a genus of rhinoceros of the tribe Elasmotheriini endemic to Asia during the Miocene living from 20—16.9 mya existing for approximately 3.1 million years.

Taxonomy

Bugtirhinus was named by Antoine and Welcomme (2000). Its type is *Bugtirhinus praecursor*. It was assigned to Elasmotheriini by Antoine and Welcomme (2000); and to Iranotheriinae by Guérin and Pickford (2003).

Source (edited): "http://en.wikipedia.org/wiki/Bugtirhinus"

Caedocyon

Caedocyon ("fit for cutting dog") is an extinct genus of bone crushing omnivorous mammal similar to a dog of the family Canidae) which inhabited western North America during the Oligocene living from 30.8—20.6 Ma and existed for approximately 10.2 million years.

Though a carnivore, dentition suggests this animal was a hypercarnivore or mesocarnivore.

Taxonomy

Caedocyon was named by Wang (1994). It is not extant. Its type is *Caedocyon tedfordi.* It was assigned to Canidae by Wang (1994) and Munthe (1998).

Like other ancient and extinct members of this subfamily, Caedocyon is a very primitive or apomorphic canid form. Its position within the evolutionary study of Hesperocyoninae is not clear. It is, however, identified by having shortened upper premolars, enlarged caniniform upper third incisors, as well as reduced upper molars. Its simple, high premolars and the precision of its occlusion suggest a relation with *Paraenhydrocyon.*

Represented by a single partial cranium.

Source (edited): "http://en.wikipedia.org/wiki/Caedocyon"

Cainotherium

Cainotherium commune is an extinct rabbit-sized herbivore that lived in Europe during the Oligocene. It is believed that the 30 cm (1 ft) long *Cainotherium* was an even-toed ungulate, usually placed in the suborder Tylopoda, along with modern camelids. It had cloven hooves, similar to those of bovids or deer, although the shape and length of its limbs suggests that it moved by leaping, like a rabbit. The shape of the teeth also suggests a rabbit-like diet, while the size of the auditory bulla and shape of the brain suggest that it would have had good senses of hearing and smell. It might have had rabbit-like ears. It had no known close relatives.

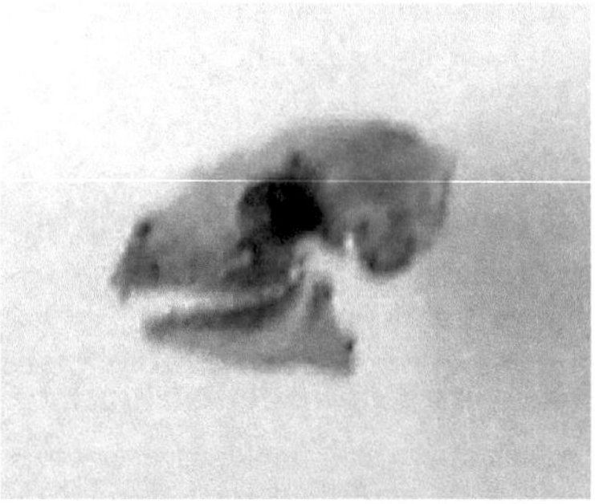

Cainotherium laticurvatum skull at the Museum für Naturkunde, Berlin

Source (edited): "http://en.wikipedia.org/wiki/Cainotherium"

Canis lepophagus

Timeline of canids highlighting *Canis lepophagus* in red. (Tedford & Wang)

Canis lepophagus or **Hare-eating Wolf** is an extinct species of canidae which was endemic to much of North America and lived from the Miocene epoch through Early Pleistocene, 10.3—1.8 Mya. The species existed for approximately 8.5 million years. It is one of the more basal species of *Canis*, having existed before most of the major clades split. It was a small, narrow skulled canid which may have given rise to grey wolves and coyotes. Some larger, broader skulled *C. lepophagus* fossils found in northern Texas may represent the ancestral stock from which true wolves derive.

Canid competitors

The Johnston's Coyote lived during a period with other canidae, specifically Borophaginae such as Epicyon (20.6—5.330 Ma), Paratomarctus (16.3—5.3 Ma), Borophagus (23.3—3.6 Ma), Carpocyon (20.4—3.9 Ma), and Aelurodon (23.03—4.9 Ma).

Fossil distribution

The Johnston's Coyote was named by Johnson in 1938. The first fossil record was found in Cita Canyon, Texas. Subsequent discoveries of specimens were found in 4 other Texas sites, Tonuco Mountain, New Mexico, western Washington Sante Fe River, Florida, Black Ranch in northern California, sites in Nebraska, Idaho, Utah, and Oklahoma.

A specimen was estimated by Legendre and Roth to weigh 18.5 kg (40.7 lbs) and another specimen was estimated to weigh 17 kg (37.4 lbs).

Source (edited): "http://en.wikipedia.org/wiki/Canis_lepophagus"

Carpocyon

Carpocyon is an extinct member of the Borophaginae, and a terrestrial canine (Bear-dog) which inhabited most of North America during the Barstovian stage of the Middle Miocene through the Hemphillian stage of the Late Miocene epoch 20.4—3.9 Mya. *Carpocyon* existed for approximately 16.5 million years.

Taxonomy

Carpocyon was named by Webb in 1969 and assigned to Canidae by Webb that same year. In 1988, Carroll assigned it to Canidae with X. Wang assigning it to Borophagina in 1999.

Morphology

Two fossil specimens of *Carpocyon* were measured by Legendre and Roth in 1988. They estimated that specimen one weighed 24.1 kg (53 lb) and the second weighed 21.8 kg (48 lb). Its physical size was between a jackal and a small wolf.

Species

- ***Carpocyon compressus*** (synonymous with *Cynodesmus cuspidatus*) lived 23.13—13.6 Ma, existing for 23.13 million years. Weight of two specimens were estimated at 19.3 kg (43 lb) and 17.6 kg (39 lb)
- ***Carpocyon limosus*** lived from 16.3—10.3 Ma, existing for 17.97 million years.
- ***Carpocyon robustus*** lived 13.6—10.3 Ma, existing for 13 million years. Weight of two specimens were 27 kg (60 lb) and 24.3 kg (54 lb).
- ***Carpocyon webbi*** lived from 16.3—10.3 Ma, existing for 6 million years. Weight of two specimens were estimated at 30.4 kg (67 lb) and 27.2 kg (60 lb)

Sister genera

Borophagus, Epicyon, Paratomarctus, Protepicyon

Source (edited): "http://en.wikipedia.org/wiki/Carpocyon"

Castor californicus

Castor californicus, is an extinct species of beaver that lived in western North America from the end of the Miocene to the early Pleistocene. The species was similar to but larger than the living species *Castor canadensis*.

Source (edited): "http://en.wikipedia.org/wiki/Castor_californicus"

Cephalogale

Cephalogale is an extinct genus of bear-like hemicyonid carnivore of the late Oligocene through Miocene epochs, endemic to North America and Europe living from around 33.9—20 Ma, existing for approximately 13.9 million years.

Before it was reconsidered to be close to the ancestry of hemicyonids, *Cephalogale* was once considered to be an ancestor of all bears.

Taxonomy

Cephalogale was named by Henri Filhol in 1879. Its type is *Cephalogale geoffroyi*. It was assigned to Ursidae by Carroll (1988); to Hemicyoninae by Hunt (1998); and to Ursoidea by Wang et al. (2005).

Fossil distribution

- Dětaň, Czech Republic ~33.9—28.4 Ma.
- Cetina de Aragon, Spain ~22.4—20 Ma.
- Standing Rock Quarry, Zia Sand Formation, Sandoval County, New Mexico ~24.8—20.6 Ma.
- Agate Springs Quarries, Sioux County, Nebraska ~23—5.3 Ma.
- Hemingford Quarry 12D, Runningwater Formation, Box Butte County, Nebraska ~20.6—16.3 Ma.

Source (edited): "http://en.wikipedia.org/wiki/Cephalogale"

Ceratotherium neumayri

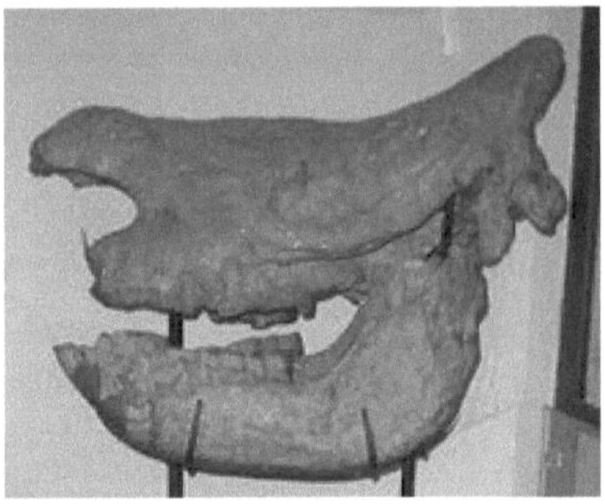

Skull

Ceratotherium neumayri is a fossil Rhinoceros from the eastern Mediterranean region dating to the late Miocene and is the likely ancestor of both the White Rhinoceros and Black Rhinoceros linneages of Africa.
Source (edited): "http://en.wikipedia.org/wiki/Ceratotherium_neumayri"

Cetotherium

Cetotherium (meaning *whale beast*) is a genus of the extinct cetaceans from the family **Cetotheriidae** (the **cetotheriids** or **cetotheres**).

Known species

To date, several whales of the genus ***Cetotherium*** have been identified; *C. furlongi*, *C. mayeri*, *C. megalophysum*, *C. parvum*, *C. priscum*, and *C. rathkii*.

Evolution

Cetotheres came in to existence during the Oligocene epoch. The cetotheres have been divided in to two sub-groups. One group includes *Cetotherium*. From evolutionary perspective, these whales share some characteristics of the Balaenopteridae and Eschrichtiidae.

Potential threats

Fossil records of Cetotheriids indicate that these whales were attacked and preyed upon by large predatory sharks (e.g. megalodon and great white sharks).
Source (edited): "http://en.wikipedia.org/wiki/Cetotherium"

Chalicotherium

Chalicotherium (Ancient Greek χαλιξ/*khalix*, *khalik-*: pebble/gravel + θηρίον/*thērion*, diminutive of θηρ/*thēr* : beast) is a genus of extinct browsing odd-toed ungulates of the order Perissodactyla and family Chalicotheriidae, endemic to Europe, Africa, and Asia during the Late Oligocene to Lower Pliocene, living from 16—7.75 mya, existing for approximately 8.25 million years.

This animal would look much like other chalicotheriid species: an odd looking herbivore with long clawed forelimbs and stouter weight bearing hindlimbs.

The type species, *Chalicotherium goldfussi*, from Miocene and Pliocene Europe, was described by J. J. Kaup in 1833 and since then 7 other species have been confidently assigned to this genus. According to current phylogenetic analyses *Chalicotherium* has two daughter genera nested within in, *Anisodon* Lartet, 1851 and *Nestoritherium* J. J. Kaup, 1859, thus rendering it paraphyletic.

Description

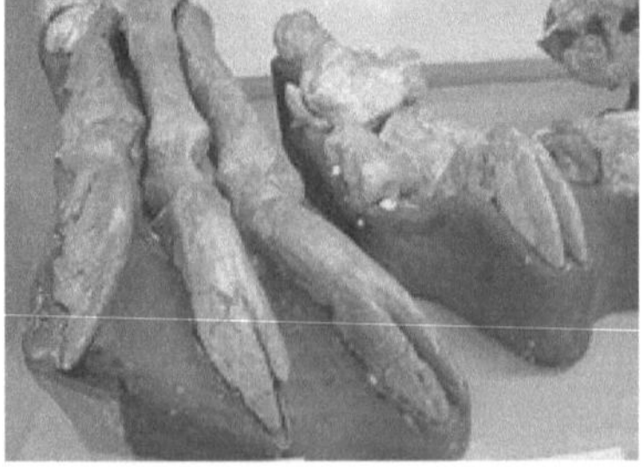

Foot bones of *Chalicotherium grande.*

Chalicotherium, like many members of Perissodactyla, was adapted to browsing, though uniquely adapted to do so among its ungulate relatives. Its arms were long and heavily clawed, allowing them to walk on their knuckles only. The arms were used to reach for high branches and bring them close to its short-faced head to strip them clean of leaves. The horse-like head itself shows adaptation to a diet of soft vegetation, since, as the animal reached sexual maturity, the incisors and upper canines were shed, suggesting that its muscular lips and the resulting gum pads were enough to crop fodder which was then processed by squarish, low-crowned molars.

Callosities on the ischium imply that these animals would sit on their haunches for extended periods of time, probably while feeding. Pad-supporting bony growth on the dorsal side of the manual phalanges is interpreted as evidence of knuckle-walking, which would probably be useful to avoid wearing down the claws, preserving them for use either as a forage collecting rake or as a formidable defensive weapon.

All of these characteristics show some convergence with the unrelated ground sloths, gorillas and giant pandas.

Classification

Taxonomic history

The type specimens for *Chalicotherium goldfussi* were found in the Upper Miocene strata of the *Dinotheriensande* beds near Eppelsheim, in the Grand Duchy of Hesse, Germany. Kaup, when describing this new animal in 1833, found the teeth to be pebble-like and named the creature accordingly. Later on, limbs found in strata located at Sansan in the department of Gers, Southwestern France, were first described as *Macrotherium* by Lartet in 1837. Further study of these fossil remains and subsequent finds by Filhol warranted a referral of the material described as *Macrotherium* to *Chalicotherium*. A more recently synonymized genus is *Butleria*, preoccupied by a butterfly genus.

Referral history for each species is detailed in the species list below along with morphological and geographical data where available.

Species

Valid:

- *Chalicotherium goldfussi* J. J. Kaup, 1833.

The type species, it was found in Upper Miocene beds located in Germany. It weighed around 1500 kg and was 2.6 m high at the shoulder.

- *Chalicotherium giganteum* Pictet, 1844.

First described as *Macrotherium giganteum* Gervais, later sunk into the type species for lack of distinguishing skeletal morphology, it was found in the Upper Miocene strata located at Sansan, France.

- *Chalicotherium brevirostris*

First described as *Macrotherium brevirostris* Colbert 1934, this species hails from the Upper Miocene Tung Gur Formation, Inner Mongolia, China.

- *Chalicotherium rusingense* Butler, 1962.

This species hails from the Lower Miocene strata located in Kenya and Uganda.

- *Chalicotherium pilgrimi*

Formerly known as *Schizotherium pilgrimi* Forster Cooper, then referred to *Macrotherium*, this species hails from the Lower Miocene Bugti beds of Pakistan.

- *Chalicotherium wetzleri*

Formerly named as *Schizotherium wetzleri* this species hails from Oligocene beds located in France and Germany, leading scientists to believe it was widespread in Western Europe.

- *Chalicotherium salinum*

First described as *Macrotherium salinum* Forster Cooper, this species was first discovered at the Lower Pliocene Lower Siwaliks beds in India; its chronological and geographic range was later extended to the Middle and Upper Miocene, and to Pakistan and China, respectively.

- *Chalicotherium wuduensis*

This species hails from Upper Miocene strata located in Gansu, China.

Invalid:

- *Chalicotherium antiquum* J. J. Kaup, 1833.

Found at the same locality as the type species, it was later found wanting of diagnostic features and sunk into the type species.

- *Chalicotherium baltavarense*

This name pertains to a specimen found in Slovakia that was later lumped into *Chalicotherium pentelicum*.

- *Chalicotherium minus*

For sometime considered a species of *Anisodon*, it was later sunk into the type species.

- *Chalicotherium posterigenium*

Junior synonym of *Chalicotherium sivalense*

- *Chalicotherium sindiense*

Junior synonym of *Chalicotherium sivalense*

- *Chalicotherium sinense*

This species hailing from Lower Pliocene beds found in China is now assigned to the genus *Nestoritherium*.

- *Chalicotherium sivalense*

Erected from specimens found in Upper Miocene beds located in India it is now assigned to the genus *Nestoritherium*.

- *Chalicotherium pentelicum* (*pentelici* in some older publications)

Hailing from Upper Miocene strata of Slovakia and Greece, being Samos Island a notable locality of the latter, the species named *Colodus pachygnathum* was later subsumed with it. Further study referred this species to the genus *Ancylotherium*.

- *Chalicotherium grande* Blainville, 1849.

Further study after this species' first description as *Macrotherium grande* Lartet, 1837, and subsequent referral to *Chalicotherium* by Blainville and later on of *M. rhodanicum*, *M. magnum* and *M. sansaniense*, warranted all the pertaining specimens to be clustered under the revived genus *Anisodon* Lartet, 1851, by Geraads *et al.* in 1995, as *Anisodon grande*.

Attributed specimens:

- *Chalicotherium* cf. *C. brevirostris* Wang *et al.*, 2001.

Hailing for the Tsaidam Basin, northern Qinghai-Tibetan Plateau, China.

- "Chalicotherium modicum" Stehlin, 1905.

A *nomen nudum*, actually a *Schizotherium priscum* tooth.

- "*Chalicotherium*" *bilobatum* Cope.

Hailing from the Oligocene of Saskatchewan, this very fragmentary specimen was the type on which Russel erected the genus *Oreinotherium*.

- *Chalicotherium* spp.

Specimens found in two Tajikistan localities, thought to pertain to at least two different species.

In popular culture

While the Chalicotherium is not well-known in popular culture, the 2001 educational documentary *Walking with Beasts* included the animal, adding some publicity.

Chalicotherium does appear briefly in the animated sequel to *Ice Age (film)*, *Ice Age: The Meltdown*.

In addition, "chalikos" are used as steeds in Julian May's Saga of Pliocene

Exile, a time travel series in which humans go through a gateway back to the Pliocene Era.
Source (edited): "http://en.wikipedia.org/wiki/Chalicotherium"

Chamitataxus

Chamitataxus is a prehistoric badger genus. *Chamitataxus avitus* is the only known species of the genus. *Chamitataxus* lived during the Late Miocene, around 6 million years ago in what is now North America. Out of the three taxidiine badgers to have existed on the continent, *Chamitataxus* is the most primitive. Very few taxidiine badger remains have been uncovered to date, with only prehistoric *Taxidea* and *Pliotaxidea* specimens being discovered prior to the *Chamitataxus* holotype being found.

Description

Chamitataxus is known only from a single holotype found in 1935, which consists of a nearly complete skull. The holotype was discovered in a quarry not far from Lyden, New Mexico, where the prehistoric herbivore *Osbornoceros* was also discovered. *Chamitataxus* was indeed named after the rock formation it was discovered in, the Chamita Formation. Due to the relative lack of *Chamitataxus* specimens, much of the research into the genus is based upon research into better represented taxa related to *Chamitataxus*, such as *Taxidea*, the modern American badger, which also existed in the Late Miocene.

Chamitataxus was fairly similar to its modern day relatives in most respects; it was a carnivore and judging from the skull found, it looked fairly similar. However, because so little of *Chamitataxus* has been found, estimating its size and other features is impossible; much is unknown about the genus. *Chamitataxus* had a very good sense of smell and had a firm grip like its modern equivalent and this enabled *Chamitataxus* to kill burrowing animals with a bite to the neck. Like other mammals, *Chamitataxus* excelled at hearing low-frequency noises, which it could hear at a long distance. Overall, *Chamitataxus* was an expert hunter based on scientist's findings, and was able to prey on many different types of land-dwelling creatures during the Miocene.
Source (edited): "http://en.wikipedia.org/wiki/Chamitataxus"

Chasicotherium

Chasicotherium rothi was a large notoungulate discovered in the Chasico Formation, in the stream homónimo of the Party of Villarino, Province of Buenos Aires, Argentina. The sediments in which the animal remains were discovered have an antiquity between 10 and 9 million years. It was an herbivore that preferred dry and open atmospheres. Like *Toxodon* and *Trigodon*, certain similarities to the hippopotamus and rhinoceros exists in its build, but there is no special relationship to either animal. This phenomenon may be an example of "adaptive convergence" or "parallel evolution", which is to say, species that are unrelated to each other evolve similarities through adapting to very similar environments and occupying equivalent ecological niches. Most notable of *Chasicotherium* was its trait that, instead of having ungulates' phalanges or hooves in his legs, it had robust claws. Its weight was approximately one ton. It was a great herbivore of the Tertiary Pampas.
Source (edited): "http://en.wikipedia.org/wiki/Chasicotherium"

Chasmaporthetes

Chasmaporthetes, also known as **Hunting** or **Running Hyena**, is an extinct genus of hyena endemic to North America, Africa, and Asia during the Pliocene-Pleistocene epochs, living from 4.9 mya—780,000 years ago, existing for approximately 4.12 million years. The genus probably arose from Eurasian Miocene hyenas such as *Thalassictis* or *Lycyaena*, with *C. borissiaki* being the oldest known representative. The species *C. ossifragus* was the only hyena to cross the Bering land bridge into the Americas, and ranged over what is now Arizona and Mexico during Blancan and early Irvingtonian Land Mammal Ages, between 5 to 1.5 million years ago.

Chasmaporthetes was one of the so called "dog-like" hyenas (of which the aardwolf is the only survivor), a hyaenid group which, in contrast to the now more common "bone-crushing" hyenas, evolved into slender limbed coursorial hunters like modern canids.

The genus has entered the popular culture lexicon as a result of cryptozoologic claims, having been proposed as the likely origin of the American Shunka Warakin and the Cuitlamiztli.

Taxonomy and etymology

Chasmaporthetes was named by Hay (1921). Its type is *Chasmaporthetes ossifragus*. It was assigned to *Hyaenidae* by Hay (1921), Geraads (1997) and Flynn (1998).

Its name means "he who saw the canyon", referring to the fact that it was the only one of its kind to cross the Bering land bridge.

Anatomy and paleoecology

The limb bones of *Chasmaporthetes* were long and slender like those of cheetahs, and its cheek teeth were slender and sharp-edged like those of a cat. It is likely that *Chasmaporthetes* probably inhabited open ground and was a daytime hunter. In Europe, the species *C. lunensis* competed with the giant cheetah *Acinonyx pardinensis*, and may have preyed on the small bourbon gazelle (*Gazella borbonica*) and the chamois antelope (*Procamptoceras brivatense*). The North American *C. ossifragus* was similar in build to *C. lunensis*, but had slightly more robust jaws and teeth. It may have preyed on the giant marmot *Paenemarmota*, and competed with the far more numerous *Borophagus diversidens*.

Source (edited): "http://en.wikipedia.org/wiki/Chasmaporthetes"

Chilecebus carrascoensis

Chilecebus carrascoensis is an extinct species of New World monkey that lived in what is now Chile during the Early Miocene some 20 million years ago. It had a body mass of about 1.3 pounds (600 grams).

Source (edited): "http://en.wikipedia.org/wiki/Chilecebus_carrascoensis"

Chilotherium

Chilotherium is an extinct genus of rhinoceros endemic to Eurasia and Asia during the Miocene through Pliocene living for 13.7—3.4 mya, existing for approximately 10.3 million years.

Taxonomy

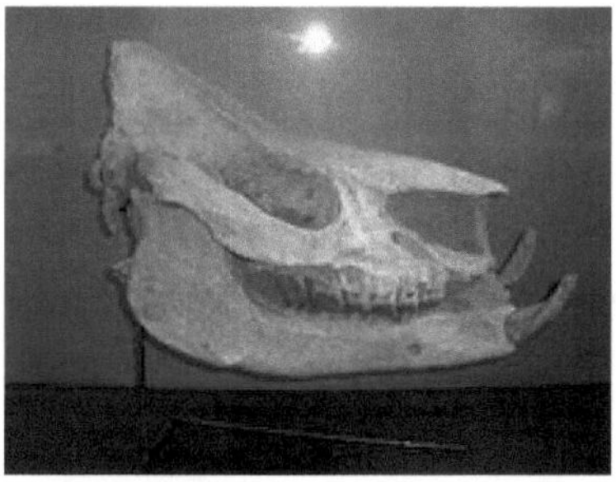

Profile of skull

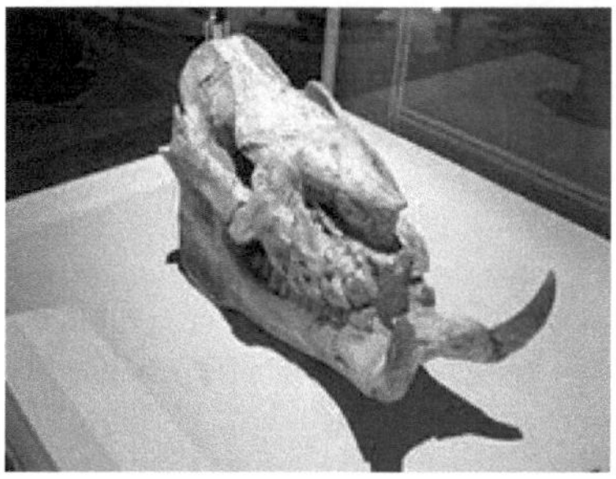

Chilotherium skull.

Chilotherium was named by Ringström (1924). It was assigned to *Rhinocerotidae* by Carroll (1988); to *Aceratheriini* by Antoine and Saraç (2005); and to *Chilotheriini* by Deng (2005).

Source (edited): "http://en.wikipedia.org/wiki/Chilotherium"

Chororapithecus abyssinicus

Chororapithecus abyssinicus was an ape that lived about 10 to 10.5 million years ago during the Miocene Epoch. It is believed to be the earliest known species of gorilla. Its existence indicates that the last common ancestor between the human/chimpanzee lineage and gorillas may have lived greater than 10 to 11 million years ago, which is at least 2 million years earlier than the previously thought date of divergence of about 8 million years ago.

The only evidence found of this extinct ape is currently nine Fossilized teeth of at least three individuals, recovered from the Chorora Formation which runs along the southern Afar Depression of Ethiopia (the same place where the remains of Lucy were discovered in 1974). Analysis of eight molars (two of them fragmented) and a canine tooth show that their structure is partly similar to modern gorillas.

The researchers compared the make up of the teeth to other current and fossil apes, and concluded that the new ape fossils possibly were a species of gorilla which ate mostly high-fiber plants, and that the fossil species is likely a 'direct ancestor' of the gorillas that currently live in Africa. Alternatively, the idea that the finds are the remains of early hominins has not been ruled out entirely.

Previous research

Previous efforts to find fossils of great ape ancestors in Africa from between 12 and 8 million years ago had been largely unsuccessful. This absence led some to hypothesize that apes were absent from Africa during this time, and had recolonized Africa from areas in Asia where fossils from this time period are more plentiful. Molecular estimates that humans and gorillas diverged after 8 million years ago were consistent with this idea, in which an adaptive radiation might have occurred after a single species entered Africa and began to adapt to different environments.

Current fossils and research prior to this finding indicated that the evolutionary split between gorillas and humans occurred around 8 million years ago. The new fossils indicate that the split may have happened as long as 10.5 million years ago. It is thought that humans shared a common ancestor with chimpanzees as recently as 4 to 7 million years ago.

Possible implications

"Based on this fossil, that means the split is much earlier than has been anticipated by the molecular evidence. That means everything has to be put back," said researcher at the Rift Valley Research Service in Ethiopia and a co-author of the study, Berhane Asfaw. This refers to the calibration of the molecular clock, the average rate of fixation of neutral mutations in the human and ape ancestors living in this environment over the past 10 million years. The human-gorilla split could only occur 10 million years ago if the mutations arose five times slower than previously thought, suggesting that other common ancestors of great apes likewise arose longer ago than believed.

Other expert opinions

Despite the finds, other researchers are not convinced that the conclusions are correct. Although the teeth are very similar to those of modern gorillas, they could have been shaped by parallel evolution of a genetically different species which consumed similar foods. "It is stretching the evidence to base a time scale for the evolution of the great apes on this new fossil. These structures appear on at least three independent lineages of apes, including gorillas, and they could relate to a dietary shift rather than indicating a new genetic trait," said Professor Peter Andrews at the Natural History Museum in London, UK, who also added, "but the fossil evidence for the evolution of our closest living relatives, the great apes, is almost non-existent."

Palaeoanthropologist Jay Kelley of the University of Illinois at Chicago was also skeptical. He remained unconvinced that it was a gorilla, saying that more work would need to be done to determine where this specimen may fit in hominid evolution. He said he would be "very cautious" about realigning divergence dates between species based on this specimen.

These teeth are collectively indistinguishable from modern gorilla subspecies in dental size and represented proportions. This modest sample nevertheless exhibits substantial size variation, with molars at both the largest and smallest end of the modern gorilla ranges of variation.

Source (edited): "http://en.wikipedia.org/wiki/Chororapithecus_abyssinicus"

Chubutherium

Chubutherium ferelloii is an extinct species of ground sloth (Cattoi, 1962).

This species of ground sloth shared many of the traits with members of the Mylodontidae family. *C. ferelloii* first appeared in the Oligocene and died out during the Middle Miocene. It is related to the Pleistocene genus, *Scelidotherium*. It possessed but a few molars, like many other sloths. Only a few specimens have been found in Patagonia.

Source (edited): "http://en.wikipedia.org/wiki/Chubutherium"

Cladosictis

Cladosictis is an extinct genus of South American marsupial.

It was an otter-like creature of about 80 cm (2 ft 8 in) long. Its body and tail were long, the limbs were short. *Cladosictis* probably hunted small creatures in low undergrowth, using its low posture for cover. Possibly, it also stole bird and reptile eggs. With sharp canines and slicing carnassials, *Cladosictis*'s teeth were similar to those of modern carnivorans, although the groups were unrelated.

Source (edited): "http://en.wikipedia.org/wiki/Cladosictis"

Climacoceras

Climacoceras (from Greek, "Ladder Horns") was a genus of early Miocene artiodactyl ungulates of Africa and Europe. The members of *Climacoceras* were related to giraffes, as the genus was once placed within Giraffidae. Fossils of the two best known species of *Climacoceras*, *C. africanus* and *C. gentryi* have been both found in Kenya. The animals measured about 1.5 m in height and had large ossicones resembling antlers. *C. africanus* had ossicones resembling tall thorn-covered plant stems, while the ossicones of *C. gentryi* resembled thorny crescents.

The genus was once placed within Palaeomerycidae, and then within Giraffidae, it is now considered a giraffoid and a new family, "Climacoceratidae," has been erected by Hamilton for this genus.

Other Climacoceratidae include *Prolibytherium* from Egypt and Libya and *Orangemeryx* from Namibia and South Africa.

Source (edited): "http://en.wikipedia.org/wiki/Climacoceras"

Climacoceratidae

Climacoceratidae ("Ladder Horns") is a family of superficially deer-like artiodactyl ungulates that were restricted to the Miocene of Africa. They are close to the ancestry of giraffes, with some genera, such as *Prolibytherium*, having originally identified as being giraffes.

The climacoceratids, namely, of what is now the type genus *Climacoceras*, were once originally placed within the family Palaeomerycidae, and then within Giraffidae. In 1978, Hamilton erected a new family, placing it close to Giraffidae.

They differ from giraffes, in that the antler-like ossicones are derived from different bones.

Source (edited): "http://en.wikipedia.org/wiki/Climacoceratidae"

Cormohipparion

The extinct ***Cormohipparion*** was originally described as a new genus of horse, and assigned to the tribe Hipparionini. However it was soon argued that the partial material fell within the range of morphological variation seen in *Hipparion*, and that the members of *Cormohipparion* belonged instead within *Hipparion*. This rested on claims that preorbital morphology did not have any taxonomic significance, a claim that detailed study of quarry sections later showed to be false. A reappraisal of many horse genera was thus conducted in 1984, and the proposed synonymy was not acknowledged by later literature.

The genus is considered to represent an ancestor to *Hippotherium*. Its fossils have been recovered from as far south as Mexico.

Source (edited): "http://en.wikipedia.org/wiki/Cormohipparion"

Cosoryx

Cosoryx furcatus is an extinct species of pronghorn that lived in the Miocene of Nevada. Not a lot is known about this animal.

Source (edited): "http://en.wikipedia.org/wiki/Cosoryx"

Cramauchenia

Cramauchenia is a member of the litoptern order of South American mammals. *Cramauchenia* was named by Florentino Ameghino. The name has no literal translation. Instead, it is an anagram of the name of a related genus *Macrauchenia*.

Source (edited): "http://en.wikipedia.org/wiki/Cramauchenia"

Cranioceras

Cranioceras is an extinct genus of artiodactyl from the Miocene to the Pliocene in the United States.

Source (edited): "http://en.wikipedia.org/wiki/Cranioceras"

Cuyamacamelus

Cuyamacamelus is an extinct genus of terrestrial herbivore the family Camelidae, endemic to North America during the Miocene 23.03—5.3 mya existing for approximately 17.73 million years.

Taxonomy

Cuyamacamelus was named by Kelly (1992). Its type is Cuyamacamelus jamesi. It was assigned to Camelidae by Kelly (1992) and Honey et al. (1998); and to Miolabinae by Whistler and Webb (2005).

Morphology

A single specimen was examined for estimated body mass by M. Mendoza, C. M. Janis, and P. Palmqvist. The specimen was estimated to weigh 485.4 kg (1,100 lb).

Source (edited): "http://en.wikipedia.

org/wiki/Cuyamacamelus"

Cynarctina

Cynarctina is an extinct hypocarnivorous canine sub-tribe clade of *Borophaginae* (Bear-dog). *Cynarctina* lived during the Hemingfordian stage of Early Miocene through Clarendonian stage of the Middle Miocene 20.6—10.3 Ma, existing for approximately 4.3 million years.

Taxonomy

Cynarctina was named by Wang et al. (1999). It was assigned to Borophagini by Wang et al. (1999).

Physical characteristics

Cynarctines feature the most dention having tubercles or rounded cusps on the molar teeth known among canids. *Metatomarctus*, *Euoplocyon*, *Psalidocyon*, *Microtomarctus*, *Protomarctus*, and *Tephrocyon* (the next series of taxa) are of medium size and occupy a sequence having projections resembling the teeth of a comb and have a rather diverse dentition. These include the most hypercarnivorous borophagine *Euoplocyon*, the peculiarly distinct *Psalidocyon*, and the dwarf lineage *Microtomarctus*.

Source (edited): "http://en.wikipedia.org/wiki/Cynarctina"

Cynarctoides

Cynarctoides is an extinct genus of terrestrial canine of the family *Borophaginae* and tribe Phlaocyonini which inhabited most of North America during the Whitneyan stage of the Early Oligocene to the Barstovian stage of the Middle Miocene 33.3—13.6 Mya existing approximately 19.7 million years.

Origin

Cynarctoides is derived from the *Nothocyon* and gave rise to *Borophaginae*, a short-face, heavy-jawed canine usually massive in size. *Cynarctoides* shared part of its period of time with a variety of bear dogs like *Aelurodontina*, *Tomarctus* and the giant mustelid genus of bone-crushing canidae.

Taxonomy

Cynarctoides was named by P. O. McGrew in 1938 and R. L. Carroll in 1988 and assigned to the family *Canidae*. Xiaoming Wang later placed it in the *Borophaginae* genus of *Phlaocyonini*.

Morphology

Fossil specimens of two individuals' body mass were examined by Legendre and Roth.

- Specimen 1 was estimated at 1.28 kg (2.8 lb).
- Specimen 2 was estimated at 1.21 kg (2.7 lb).

Species

Species within *Cynarctoides* are:

- *Cynarctoides acridens* (synonymous with *Cynarctus mustelinus*) existed for 15 mya †
- *Cynarctoides emryi* existed for 4.3 mya †
- *Cynarctoides gawnae* existed for 15 mya †
- *Cynarctoides harlowi* existed for 4 mya †
- *Cynarctoides lemur* existed for 11.3 mya †
- *Cynarctoides luskensis* existed for 4. 2 mya †
- *Cynarctoides roii* existed for 4.5 mya †

Fossil distribution

Sites representing each species of *Cynarctoides* with age of find:

- Sharps Site, Sharps Formation, Shannon County, South Dakota, (C. roii) ~30.8—26.3 Ma.
- Buda Mine Site, Alachua County, Florida (C. lemur) ~24.8—20.6 Ma.
- Chugwater Site, Platte County, Wyoming (C. luskensis) ~24.8—20.6 Ma.
- Van Tassel, Niobrara County, Wyoming (C. harlowi) ~24.8—20.6 Ma.
- Split Rock Site, Arikaree Formation, Natrona County, Wyoming, (C. acridens) ~20.6—16.3 Ma.
- Hemingford Quarry 7B, Runningwater Formation, Box Butte County, Nebraska (C. emryi) ~20.6—16.3 Ma.
- Blickomlyus Hill Site, Zia Sand Formation, Sandoval County, New Mexico (C. gawnae) ~20.3—5.3 Ma.
- Hidalgo Bluff Site, Oakville Formation, Washington County, Texas (C. acridens) ~20.3—5.3 Ma.

Source (edited): "http://en.wikipedia.org/wiki/Cynarctoides"

Cynarctoides acridens

Cynarctoides acridens is an extinct species of Cynarctoides, belonging to the subfamily Borophaginae and tribe Phlaocyonini, a canine which inhabited most of west central North America from the Miocene epoch (Harrisonian stage to Early Barstovian stages) living 24.8—13.6 mya and existed for approximately 15.6 million years.

Description

Borophaginae were short-face, heavy-jawed canines. *Cynarctoides acridens* shared part of its period of time with a variety of bear dogs like *Aelurodontina*, *Tomarctus* and the giant mustelid genus of bone-crushing canidae.

Origin

Cynarctoides acridens is synonymous with *Cynarctus mustelinus* and was named by E. H. Barbour and H. J. Cook in 1914. Its discovery was in a Harrisonian terrestrial horizon in the Upper Harrison Beds Formation of Nebraska. *Cynarctoides acridens* was recombined as *Notharctus acridens* by William Diller Matthew in 1932. Under subsequent study, it was recombined as *Cynarctoides acridens* by P. O. McGrew in 1938 followed by Galbreath in 1956, J. Munthe in 1988, and Xiaoming Wang in 1999.

Morphology

Two specimens were examined by Legendre and Roth for body mass.

- Specimen 1 was estimated to weigh 1.42 kg (3.1 lb).
- Specimen 2 was estimated to weigh 1.48 kg (3.3 lb).

Fossil distribution

- Hidalgo Bluff Site, Oakville Formation, Washington County, Texas ~23.0—5.3 Ma.
- Nambe Site, Tesuque Formation, Santa Fe County, New Mexico ~23.0—5.3 Ma.
- Split Rock, Arikaree Formation, Natrona County, Wyoming, ~20.6—16.3 Ma.
- Yermo Site, Barstow Formation, San Bernardino County, California ~16.3—13.6 Ma.

Source (edited): "http://en.wikipedia.org/wiki/Cynarctoides_acridens"

Cynarctoides emryi

Cynarctoides emryi is an extinct species of Cynarctoides, belonging to the subfamily Borophaginae and tribe Phlaocyonini, a canid which inhabited most of west central North America from the Miocene epoch (Hemingfordian stage) living 23.0–5.3 mya and existed for approximately 17.7 million years.

Taxonomy

Cynarctoides emryi was named by Xiaoming Wang et al. (1999). It was discovered at the Hemingford Quarry 7B site, which is in a Hemingfordian terrestrial horizon in the Runningwater Formation.

Morphology

Body mass

Two specimens were examined by Legendre and Roth for body mass.

- Specimen 1 was estimated to weigh 1.29 kg (2.8 lb).
- Specimen 2 was estimated to weigh 1.36 kg (3.0 lb).

Fossil distribution

- Hemingford Quarry 7B, Runningwater Formation, Box Butte County, Nebraska ~23.0–5.3 Ma.
- Wood's Canyon Quarry, Runningwater Formation, Dawes County, Nebraska ~20.6–16.3 Ma.

Source (edited): "http://en.wikipedia.org/wiki/Cynarctoides_emryi"

Cynarctoides gawnae

Cynarctoides gawnae is an extinct species of Cynarctoides, belonging to the subfamily Borophaginae and tribe Phlaocyonini, a canid which inhabited most of west central North America from the Miocene epoch (Hemingfordian stage) living 23.0—5.3 mya and existed for approximately 17.7 million years.

Taxonomy

Cynarctoides gawnae was named by Xiaoming Wang et al. (1999). Its type locality is Jeep Quarry, which is in a Hemingfordian terrestrial horizon in the Zia Sand Formation of New Mexico.

Morphology

Body mass

Two specimens were examined by Legendre and Roth for body mass.

- Specimen 1 was estimated to weigh 1.02 kg (2.2 lb).
- Specimen 2 was estimated to weigh 1.09 kg (2.4 lb).

Fossil distribution

- Blickomlyus Hill, Zia Sand Formation, Sandoval County, New Mexico ~23.0—5.3 Ma.
- Jeep Quarry, Zia Sand Formation, Sandoval County, New Mexico ~20.6—16.3 Ma.

Source (edited): "http://en.wikipedia.org/wiki/Cynarctoides_gawnae"

Cynarctoides harlowi

Cynarctoides harlowi is an extinct species of Cynarctoides, belonging to the subfamily Borophaginae and tribe Phlaocyonini, a canid which inhabited west central North America during the Late Oligocene to Miocene epoch living 24.8—20.6 mya and existed for approximately 4.2 million years.

Taxonomy

Cynarctoides harlowi was originally named *Pachycynodon harlowi* by Loomis (1932). Its type locality is Van

Tassel (Upper), which is in a Harrisonian terrestrial horizon in Wyoming. It was recombined as Nothocyon harlowi by Macdonald (1963); it was recombined as *Cynarctoides harlowi* by Xiaoming Wang, Richard H. Tedford et al. (1999).

Morphology

Body mass

Two specimens were examined by Legendre and Roth for body mass.

- Specimen 1 was estimated to weigh 1.21 kg (2.7 lb).
- Specimen 2 was estimated to weigh 1.28 kg (2.8 lb).

Fossil distribution

- Van Tassel, Niobrara County, Wyoming ~23.0—5.3 Ma.

Source (edited): "http://en.wikipedia.org/wiki/Cynarctoides_harlowi"

Cynarctoides lemur

Cynarctoides lemur is an extinct species of Cynarctoides, belonging to the subfamily Borophaginae and tribe Phlaocyonini, a canid endemic to various parts of central North America from the Late Oligocene to Miocene epoch (Harrisonian stage) living 24.8—20.6 mya and existed for approximately 4.2 million years.

Fossil evidence suggests that *C. lemur* migrated to the east and southeast over time becoming extinct in the northwest followed by upper Great Plains and then the southeast.

Taxonomy

Cynarctoides lemur was named by Cope (1879). It was recombined as Galecynus lemur by Cope (1881); it was recombined as *Cynodictis lemur* by Scott (1898); it was recombined as Nothocyon lemur by Matthew (1899), Wortman and Matthew (1899), Wortman and Matthew (1899), Merriam (1906), Matthew (1907), Thorpe (1922), Hall and Martin (1930), Macdonald (1963), Macdonald (1970) and Munthe (1998); it was recombined as Cormocyon lemur by Wang and Fremd (1994); it was recombined as *Cynarctoides lemur* by Xiaoming Wang, Richard H. Tedford, et al. (1999).

Morphology

Body mass

Two specimens were examined by Legendre and Roth for body mass.

- Specimen 1 was estimated to weigh 0.960 kg (2.1 lb).
- Specimen 2 was estimated to weigh 1.02 kg (2.2 lb).

Fossil distribution

- North Blue Basin, John Day Formation, Grant County, Oregon ~33.3—30.8 Ma.
- Sharps Site, Shannon County, South Dakota ~30.8—26.3 Ma.
- Godsell Ranch, Sharps Formation, Shannon County, South Dakota ~30.8—20.6 Ma.
- Buda Mine, Alachua County, Florida ~24.8—20.6 Ma.

Source (edited): "http://en.wikipedia.org/wiki/Cynarctoides_lemur"

Cynarctoides luskensis

Cynarctoides luskensis is an extinct species of Cynarctoides, belonging to the subfamily Borophaginae and tribe Phlaocyonini, a canid endemic to west central North America from the Late Oligocene to Miocene epoch living 24.8—20.6 Ma and existed for approximately 4.2 million years.

Taxonomy

Cynarctoides luskensis was named by Xiaoming Wang, Richard H. Tedford, et al. (1999). Its type locality is Eighteen Mile District, which is in a Harrisonian terrestrial horizon in Wyoming.

Morphology

Body mass

Two specimens were examined by Legendre and Roth for body mass.

- Specimen 1 was estimated to weigh 0.971 kg (2.1 lb).
- Specimen 2 was estimated to weigh 1.03 kg (2.3 lb).

Fossil distribution

- Eighteen Mile District, Goshen County, Wyoming ~24.8—20.6 Ma.
- Jay Em Site, Goshen County, Wyoming ~24.8—20.6 Ma.
- Chugwater Site, Platte County, Wyoming ~24.8—20.6 Ma.

Source (edited): "http://en.wikipedia.org/wiki/Cynarctoides_luskensis"

Cynarctoides roii

Cynarctoides roii is an extinct species of Cynarctoides, belonging to the subfamily Borophaginae and tribe Phlaocyonini, a canid which inhabited west central North America from the Late Oligocene to Miocene living 30.8—20.6 mya and existed for approximately 10.2 million years.

Taxonomy

Cynarctoides roii was named by Macdonald (1963). It was recombined as *Cormocyon roii* by Xiaoming Wang and Richard H. Tedford (1992) and Wang and Fremd (1994); it was recombined as *Cynarctoides roii* by Wang et al. (1999).

Morphology

Body mass

Two specimens were examined by Legendre and Roth for body mass.

- Specimen 1 was estimated to weigh 0.644 kg (1.4 lb).
- Specimen 2 was estimated to weigh 0.700 kg (1.5 lb).

Fossil distribution

- Sharps Site, Shannon County, South Dakota ~30.8—20.6 Ma.
- Horse Creek, Banner County, Nebraska ~30.8—20.6 Ma.

Source (edited): "http://en.wikipedia.org/wiki/Cynarctoides_roii"

Cynarctus

Cynarctus is an extinct canid of the subfamily Borophaginae which inhabited most of North America during the Late Miocene subepoch (15.9—10.3 mya). *Cynarctus* existed for approximately 5.6 million years.

Taxonomy

Cynarctus was named by Matthew in 1902. It was assigned to *Amphicyoninae* by Matthew (1902); to Canidae by Carroll (1988); and to *Cynarctina* by Wang et al. (1999).

Morphology

Two specimens were examined by Legendre and Roth for body mass. The first specimen was estimated to weigh 5.28 kg (11.28 lbs). The second specimen was estimated to weigh 5.16 kg (11.3 lbs).

Fossil distribution

Fossils have been uncovered in California, Maryland, western Nebraska, and Texas.

Source (edited): "http://en.wikipedia.org/wiki/Cynarctus"

Cynelos

Cynelos is a large extinct genus of terrestrial carnivores belonging to the suborder Caniformia, family Amphicyonidae ("bear dog"), and which inhabited North America, Europe Asia, and Africa from the Early Miocene subepoch to the Late Miocene subepoch 24.8—7.2 Mya, existing for approximately 17.6 million years.

Dentition of this animal displayed only the characteristics of a carnivore.

Taxonomy

Cynelos was named by Jourdan in 1862 as type European *C. lemanensis* (Europe). Its type is *Amphicyon lemanensis*. It was assigned to Amphicyoninae by Hunt (1998) and Sach and Heizmann (2001); and to *Amphicyonidae* by Jourdan (1862), Carroll (1988) and Morlo et al. (2007).

Morphology

Body mass

A single specimen was examined by Legendre and Roth in 1988 for body mass. That specimen was estimated to weigh 84.4 kg (190 lb).

Fossil distribution

Bhagothoro, Pakistan; West Turkana, Kenya; Uganda, Wadi Moghra, Egypt; Skull Spring, Oregon; Gilchrist County, Florida (Thomas Farm), Pollack Farm, Delaware.

Species

C. caroniavorus (synonymous with *Parictis bathygenus*), *C. euryodon*, *C. idoneus*, *C. lemanensis*, *C. macrodon*, *C. sinapius* (synonymous with Amphicyon amnicola)

Sister genera

Amphicyon, Ischyrocyon (synonymous with *Hadrocyon*), *Pliocyon*, *Pseudocyon* (synonymous with *Amphicyonopsis*), *Ysengrinia.*

Source (edited): "http://en.wikipedia.org/wiki/Cynelos"

Cynodesmus

Cynodesmus is an extinct genus of omnivorous canine which inhabited North America during the Oligocene living from 33.3—26.3 Ma and existed for approximately 7 million years.

Cynodemus was one of the first canids to truly look dog-like. At around 1 metre (3.3 ft) in length, it was about the same size as a modern coyote, but had a shorter skull, heavier tail, and longer rump. The shape of its limbs suggests that *Cynodesmus* was not a very good runner compared to most other canids; it probably attacked prey from an ambush. Unlike modern dogs, it had five toes on each foot, bearing partially retractable claws.

Taxonomy

Cynodesmus once included numerous species of Oligocene and Miocene canid with highly carnivorous (hypercarnivorous) dentitions. A revision of the genus by Wang (1994) indicates that most species previously placed in *Cynodesmus* are unrelated to the type species, *C. thooides*. These other species have been placed the genera *Carpocyon*, *Desmocyon*, *Leptocyon*, *Metatomarctus*, *Osbornodon*, *Otarocyon*, *Paracynarctus*, *Paratomarctus*, and *Phlaocyon* (Wang, 1994; Wang et al., 1999). Of these, only *Osbornodon* belongs in the same subfamily as *Cynodesmus*, Hesperocyoninae. The remaining genera are placed in the subfamilies Borophaginae and Caninae.

With unrelated species removed, *Cynodesmus* is currently restricted to the type species and the closely related *C. martini* (Wang, 1994). Studies using the old conception of *Cynodesmus* considered it to be the ancestor of *Tomarctus* (16-23 Ma) from which wolves, dogs, foxes and fennecs developed. The *Cynodesmus* is a good example of convergent evolution because of other species such as the *Borophagus*, the largest and most dominant canids of this Pliocene epoch, both of which evolved from it.

Source (edited): "http://en.wikipedia.org/wiki/Cynodesmus"

Cyonasua

Cyonasua is an extinct procyonid genus from the late Miocene of South America (7.5 to 5 million years ago). Its name in Greek means *dog-coati* because its features resemble those of a dog and a coati. It's ancestors arrived from Central America by island hopping, as perhaps the earliest southward mammalian migrants of the Great American Interchange. Its predators were terror birds and carnivorous metatherians. It evolved into *Amphinasua* and the bear-like *Chapalmalania*.

Source (edited): "http://en.wikipedia.org/wiki/Cyonasua"

Daeodon

Daeodon (formerly *Dinohyus*, "terrible hog"), one of the largest, if not the largest, entelodont artiodactyls, lived 25-18 million years ago in North America.

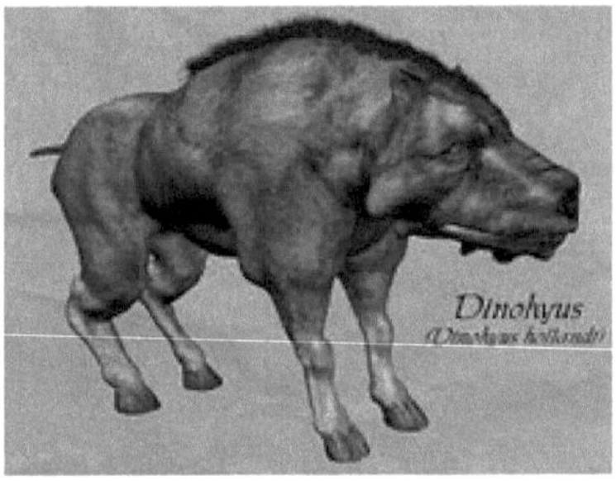

3D reconstruction of *Daeodon*

The 3.6 m (12 ft) long, 2.1 m (7 ft) tall (at-shoulder), 1 m long skull, 1000 kg mass animal strongly resembled a giant, monstrous pig or warthog, possessing huge jaws with prominent tusks and flaring cheekbones. It was a huge, bone-crushing scavenger and predator, found at Agate Springs Quarry. It had long skull bones under its eyes and bony protrusions on the lower jaw, not dissimilar to the 'warts' of the warthog, which may have supported jaw muscles. The well-known genus *Dinohyus* ("terrible pig") has now been synonymized with *Daeodon*, as the latter is the earlier name, having priority.

Source (edited): "http://en.wikipedia.org/wiki/Daeodon"

Daphoenus socialis

Daphoenus socialis is an extinct species of Amphicyonidae ("bear dogs"), a family of terrestrial carnivores belonging to the suborder Caniformia, which inhabited North America from the Miocene ~25.4—24.3 Mya, existing for approximately 1.1 million years.

Taxonomy

Daphoenus socialis was originally desinagnated as Pericyon socialis and named by Thorpe (1922) a genotype. It is the type species of Pericyon. It was recombined as Daphoenus socialis by Hunt in 1998.

Morphology

A single specimen was measured by Legendre and Roth in 1988 for estimated body mass.

- 13 kg (29 lb).

Fossil distribution

Daphoenus socialis fossils found in Oligocene Hemingfordian rocks at the Haystack Member, Wheeler County, Oregon are dated at ~24.3 Ma. and Kimberly Member, Grant County, Oregon with several other species of mammal such as Hesperocyon, Hypertragulus, and Leptomeryx dating ~25.4—25.3 Ma.

Source (edited): "http://en.wikipedia.org/wiki/Daphoenus_socialis"

Darocasorex

Darocasorex is an extinct shrew which existed in Europe during the Miocene period. It was first named by Jan A. van Dam in 2010.

Source (edited): "http://en.wikipedia.org/wiki/Darocasorex"

Deinogalerix

Deinogalerix (from Ancient Greek, "terrible/terror" + "shrew"), was a genus of the order Erinaceomorpha, which lived in Italy in the Late Miocene. The genus was apparently endemic to what was then Gargano Island, today's Gargano peninsula. The *Deinogalerix* remains were first described in 1972.

The genus belonged to the subfamily of gymnures or moon-rats, which are not rats at all, but rather hairy, superficially rat-like relatives of the hedgehog lacking quills. *Deinogalerix* had a long, thin, conical face, small pointed ears, a lengthy, tapering tail and long hairs.

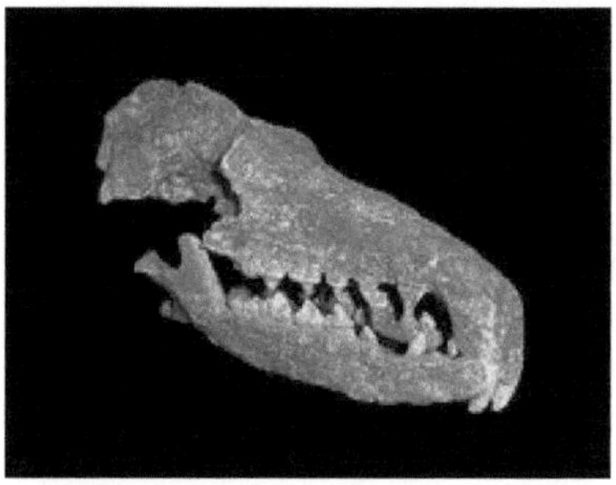

Fossil head of *Deinogalerix koenigswaldi*

reconstruction of *Deinogalerix koenigswaldi*

Deinogalerix koenigswaldi's skull was 20 cm long, and the entire body measured 60 cm. It occupied the same ecological niche as dogs and cats today, except that it had more predators itself - such as the enormous barn owl *Tyto gigantea.*

It is believed that the species of *Deinogalerix* were insectivores, mostly feeding off invertebrates like beetles, dragonflies and crickets, and possibly even snails. But the larger species may also have hunted small mammals, reptiles and birds.

In the Late Miocene subepoch, what is at present Italy was mainly a group of small islands and only at a later date did majority of these join with the mainland. It is known that animals on these islands sometimes evolved quite differently than they would elsewhere. So it is possible that *Deinogalerix* may have lived exclusively in the Gargano.
Source (edited): "http://en.wikipedia.org/wiki/Deinogalerix"

Deinotherium

Deinotherium ("terrible beast"), also called the **Hoe tusker**, was a gigantic prehistoric relative of modern-day elephants that appeared in the Middle Miocene and continued until the Early Pleistocene. During that time it changed very little. In life it probably resembled modern elephants, except that its trunk was shorter, and it had downward curving tusks attached to the lower jaw.

Deinotherium is the third largest land mammal known to have existed; only *Paraceratherium* and *Mammuthus sungari* were larger, although *Mammuthus imperator* may have rivaled it in size. Males were generally between 3.5 and 4.5 meters (12 and 15 feet) tall at the shoulders although large specimens may have been up to 5 m (16 ft). Their weight is estimated to have been between 5 and 10 tonnes (5.5 and 11 US Standard tons), with the largest males weighing in excess of 14 tonnes (15.4 US Standard tons). *Deinotherium's* range covered parts of Asia, Africa, and Europe. Adrienne Mayor, in *The First Fossil Hunters: Paleontology In Greek and Roman Times,* has suggested that deinothere fossils found in Greece helped generate myths of archaic giant beings. A tooth of a deinothere found on the island of Crete, in shallow marine sediments of the Miocene suggests that Crete was closer or connected to the mainland during the Messinian Salinity Crisis.

Evolutionary Relationships

Deinotherium giganteum

Deinotherium is the type genus of the family Deinotheriidae, evolving from the smaller, early Miocene *Prodeinotherium*. These proboscideans represent a totally distinct line of evolutionary descent to that of other elephants, one that probably diverged very early in the history of the group as a whole. The large group to which elephants belong formerly contained several other related groups: besides the deinotheres there were the gomphotheres (some of which had shovel-like lower front teeth), and the mastodons. Only elephants survive today.

Paleoecology

The way *Deinotherium* used its curious tusks has been much debated. It may have rooted in soil for underground plant parts like roots and tubers, pulled down branches to snap them and reach leaves, or stripped soft bark from tree trunks. *Deinotherium* fossils have been uncovered at several of the African sites where remains of prehistoric hominid relatives of modern humans have also been found.

Characteristics

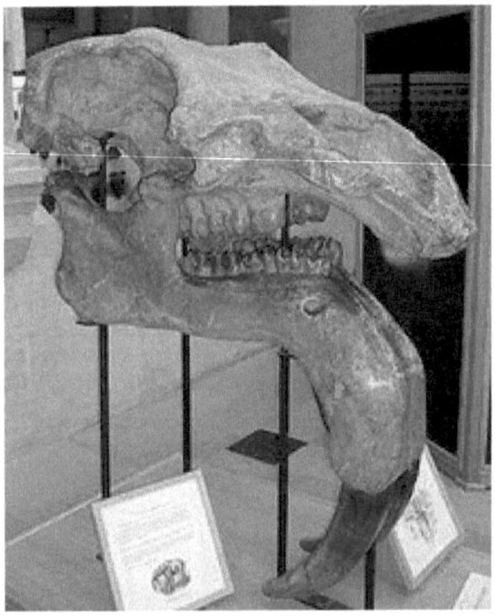

Deinotherium skull from Oxford University Museum of Natural History

The following description is for *D. giganteus* but in general applies to the other two species as well.

Permanent tooth formula 0-0-2-3/1-0-2-3 (deciduous 0-0-3/1-0-3), with vertical cheek tooth replacement. Two sets of bilophodont and trilophodont teeth. Molars and rear premolars tapiroid, vertical shearing teeth, and show that deinotheres became an independent evolutionary branch very early on; other premolars used for crushing. The cranium is short, low, and flattened on the top (in contrast to more advanced proboscids, which have a higher and more domed forehead; the implication may be that deinotheres were less intelligent than other proboscids), with very large, elevated occipital condyles. The nasal opening is retracted and large, indicating a large trunk. The rostrum is long and the rostral fossa broad. Mandibular symphyses (the lower jawbone) is very long and curved downward, which, with the backward curved tusks, is a distinguishing feature of the group; it possessed no upper tusks.

Mounted *Deinotherium* skelton

Deinotherium is distinguished from its predecessor *Prodeinotherium* by its much greater size, greater crown dimensions, and reduced development of posterior cingula ornamentation in the second and third molar.

Three species are recognized, all of great size.

Deinotherium giganteum **Kaup 1829**

Head and teeth of *Deinotherium giganteum*

Deinotherium giganteum is the type species, and is described above. It is primarily a late Miocene species, most common in Europe, and is the only species known from the circum-Mediterranean. Its last reported occurrence is from the middle Pliocene of Romania (2 to 4 million BP). The "Grigore Antipa" museum of natural history in Bucharest, Romania, has the only complete skeleton of Deinotherium giganteum in the world. It was unearthed in 1894, in Vaslui county, by the Romanian paleontologist Gregoriu Stefanescu.

Deinotherium giganteum

An entire skull, found in the Lower Pliocene beds of Eppelsheim, Hesse-

Darmstadt in 1836, measured 4 ft (1.2 m) in length and 3 ft (.9 meters) in breadth, indicating an animal exceeding modern elephants in size.

Deinotherium indicum Falconer 1945

Deinotherium indicum is the Asian species, known from India and Pakistan. It is distinguished by a more robust dentition and p4-m3 intravalley tubercles. *D. indicum* appears in the middle Miocene, and is most common in the late Miocene. It disappears from the fossil record about 7 million years BP (late Miocene).

Deinotherium bozasi Arambourg 1934

Deinotherium bozasi is the African species. It is characterized by a narrower rostral trough and smaller but higher nasal aperture, and a higher and narrower cranium, and shorter mandibular symphysis, than the other two species. *D. bozasi* appears at the beginning of the late Miocene, and continues there after the other two species have died out elsewhere. The youngest fossils are from the Kanjera Formation, Kenya, about a million years old (early Pleistocene)

Deinotherium in popular culture

Deinotherium was shown in the fourth episode of BBC's Walking with Beasts. The animal was shown to live side by side with the early human ancestor *Australopithecus*, the Chalicothere *Ancylotherium*, and the primitive cat *Dinofelis* in Ethiopia about 3.2 million years ago.

Deinotherium was also depicted in Karel Zeman's live action/stop motion film Journey to the Beginning of Time. One of the characters in the film initially mistakes it for an elephant.
Source (edited): "http://en.wikipedia.org/wiki/Deinotherium"

Dendropithecus

Dendropithecus is an extinct genus of apes native to East Africa between 20 and 15 million years ago. It may have been the ancestor of modern gibbons, which it resembled in some respects.

Dendropithecus was a slender ape, about 60 centimetres (2.0 ft) in body length. The structure of its arms suggest that it would have been able to brachiate, swinging between trees by its arms, but that it would not have been as efficient at this form of movement as modern gibbons. However, its teeth suggest a very gibbon-like diet, likely consisting of fruit, soft leaves, and flowers.
Source (edited): "http://en.wikipedia.org/wiki/Dendropithecus"

Desmatophoca

Desmatophoca is an extinct genus of pinniped.

This poorly known fossil pinniped is a member of the true seal-like extinct pinniped family, the Desmatophocidae. Unlike modern true seals, it had a tail, although this was relatively short. It had large eyes, and presumably hunted by sight, and its ears do not appear to have been adapted for hearing underwater.

Desmatophoca oregonensis was one of the first known fossil pinnipeds from the west coast of North America, and it is now known by several skulls, jaws, and forelimb elements from the Oregon coast. It had lost the primitive shearing carnassials of older pinnipeds like *Enaliarctos*, which indicates that it was able to feed and ingest fish underwater without having to return to land.
Source (edited): "http://en.wikipedia.org/wiki/Desmatophoca"

Desmocyon

Desmocyon is an extinct genus of the *Borophaginae* and a terrestrial canine which inhabited most of North America during the Harrisonian stage of the Late Oligocene through Early Hemingfordian stage of the Early Miocene epoch living 24.8—16.3 Mya existing for approximately 8.5 mya.

Fossil record

It is a rarely found genus, with fossil deposits only occurring in western Nebraska, Wyoming, New Mexico and north Florida. *Desmocyon* contains two species: *Desmocyon matthewi* and *Desmocyon thomsoni*. It was first named by Wang in 1999.

Morphology

Fossil specimens of two individuals' body mass were examined by Legendre and Roth. The first specimen was estimated to weigh 5.67 kg (13 lb). The second specimen was estimated to weigh 5.52 kg (12 lb).

Sister genera

Cormocyon, *Euoplocyon*, *Metatomarctus*, *Microtomarctus*, *Protomarctus*, *Psalidocyon*, and *Tephrocyon*
Source (edited): "http://en.wikipedia.org/wiki/Desmocyon"

Desmostylus

Desmostylus is a monotypic extinct genus of herbivorous mammal of the family *Desmostylidae* living from the Chattian stage of the Late Oligocene subepoch through the Late Miocene subepoch (28.4 mya—7.250 Mya) and in existence for approximately 21.2 million years.

Morphology

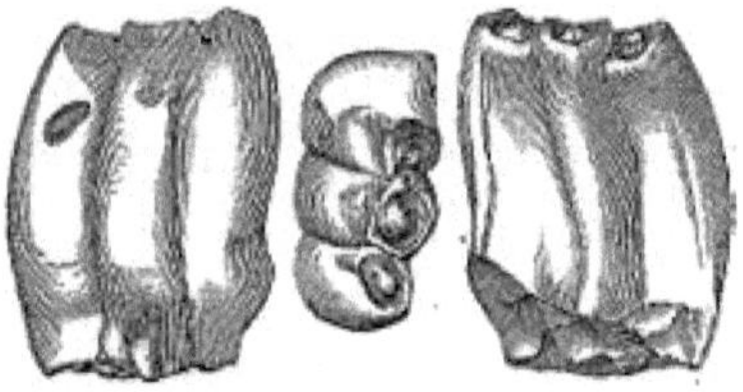

Illustration of a tooth by Marsh

Desmostylus was a large, hippopotamus-like creature of about 1.8 metres (6 ft) long which weighed about 200 kilograms (440 lb). It had a short tail and powerful legs with four hooves. Both the creature's jaws were elongated and sported forward-facing tusks, which were elongated canines and incisors.

Most likely amphibious, *Desmostylus* is thought to have lived in shallow water in coastal regions. Some paleontologists have argued it may have fed on seaweed during low tide. However, recent isotope work indicates that *Demostylus* more likely lived in freshwater or estuary ecosystems and ate aquatic freshwater plants.

Species

D. hesperus (syn. *D. japonicus, D. watasei, D. cymatias, D. californicus, D. mirabilis, D. minor, Desmostylella typica*)

Sister genera

Behemotops, Cornwallius, Kronokotherium, Paleoparadoxia, Vanderhoofius

Fossil distribution

Fossils have been discovered from along the northern Pacific Rim from Baja, Mexico northward along the coast of California, Oregon, Washington and west to Sakhalin Island, Hokkaido, Japan, and south to the Shimane Prefecture, Japan.

An artist's rendering of *Desmostylus hesperus*.

Source (edited): "http://en.wikipedia.org/wiki/Desmostylus"

Diatomys

Diatomys is a hystricomorphous sciurognathous rodent known from Miocene deposits in China, Japan, Pakistan, and Thailand. The fossil range is from the late Early Miocene to the Middle Miocene (22.5-11 Ma).

Specimens

Specifically the strata and regions from which *Diatomys* has been collected are: Shanwang series in Shandong province, China, Jiangsu province in China, Kyūshū in Japan, the Siwaliks in northern Pakistan, and Li Basin in Lamphun Province, Thailand.

Li (1974) described ***Diatomys shantungensis*** on the basis of two moderately complete specimens from Shandong. This material had good preservation of dental characters, but much of the skull was difficult to interpret due to flattening. Dawson *et al.* (2006) reported the finding of another *D. shantungensis* fossil from Shandong that showed much improved preservation of cranial and skeletal characters. Impressions of hair and whiskers were observable in the specimen.

Mein and Ginsburg (1985) described ***Diatomys liensis*** from Thailand, and considered it distinct enough to warrant recognition as separate from *D. shantungensis. D. liensis* was found to be smaller overall, with a higher molar crown height and distinctions in cusp morphology in the premolar and first molar.

Characteristics

Head and body was about 25 cm (9.8 in). The overall appearance of *Diatomys* was thought to be very similar to the extant Laotian Rock Rat.

The dental formula for *Diatomys* is 1/1 I, 0/0 C, 1/1 P, 3/3 M = 20. Incisors display multiserial enamel. Structure of the Hunter-Schreger bands is very similar to *Laonastes*. The root of the incisor is shortened. The cheek teeth are bilophodont, displaying two transverse ridges that are each slightly curved into a mild horseshoe shape. Cheek teeth have four roots on both the upper and lower jaws.

The infraorbital canal is enlarged, presumably allowing for the passage of the medial masseter muscle as with other hystricomorphs. The mandible is sciurognathous. As with other diatomyids, *Diatomys* has no coronoid process and the masseteric fossa extends far forward to below the fourth premolar.

The postcranial skeleton does not appear to have any unique specializations that might be associated with a fossorial, arboreal, or saltatorial way of life. This has led researchers to assume *Diatomys* was terrestrial.

Relationship

Several authors have suggested that *Diatomys* might belong to the family

Pedetidae (McKenna and Bell, 1997) or be related to the pedetids (Mein and Ginsburg, 1997; Marivaux *et al.*, 2004). Li (1974) also noted the possibility that it was related to geomyoids based on the bilophodont tooth. Dawson *et al.* (2006) determined that the closest known relative of *Diatomys* is the extant Laotian Rock Rat, a unique rodent first described in 2005 (Jenkins *et al.*, 2004) from Khammouan, Laos. The remaining diatomyid in their analysis, *Fallomus*, was determined to be more distant. Source (edited): "http://en.wikipedia.org/wiki/Diatomys"

Diceratherium

Diceratherium is an extinct genus of rhinoceros endemic to North America, Europe, and Asia during the Oligocene through Miocene living from 33.9—11.6 mya, existing for approximately 22.3 million years.

Taxonomy

Restoration

Diceratherium was named by Marsh (1875). It is not extant. Its type is *Diceratherium armatum*. It was assigned to Rhinocerotidae by Marsh (1875) and Carroll (1988); to Diceratheriinae by Prothero (1998); to Aceratheriinae by Weidmann and Ginsburg (1999); and to Teleoceratini by Sach and Heizmann (2001). *Diceratherium* had a two horns side-by-side on it nose. It lived during the Miocene Epoch.
Source (edited): "http://en.wikipedia.org/wiki/Diceratherium"

Dimylus

Dimylus is an extinct genus of insectivore mammal.

The creature probably resembled the modern desman in terms of size (10–20 cm in length) and physical appearance, possessing a proboscis. Its knobby teeth were small (no longer than 3,6 mm) and covered with enamel. Coupled with powerful jaw muscles this made *Dimylus* capable of crushing armored creatures such as crustaceans. *Dimylus* has been found in Europe, in areas that were abundant with water in the time it existed, suggesting it filled the same ecological niche as the modern desmans.

Source (edited): "http://en.wikipedia.org/wiki/Dimylus"

Dinaelurus

Dinictis is an extinct genus of the family *Nimravidae*, endemic to North America during the Eocene-Oligocene epochs (30.8—20.6 mya), existing for approximately 10.2 million years.

It is believed that *Dinaelurus* was a cursorial predator meaning it ran down its prey. This is suggested by the nimravid's short face and large nostril, similar to a cheetah which is also a cursorial predator.

Taxonomy

Dinaelurus was named by Eaton (1922). Its type is *Dinaelurus crassus*. It was assigned to *Nimravinae* by Flynn and Galiano (1982) and Bryant (1991); and to *Nimravidae* by Eaton (1922) and Martin (1998).

In popular culture

Dinaelurus crassus is the ancestor of the fictional species, *Dinaelurus illumina sapiens*, in the Ratha or "Named" series by author Clare Bell.

Fossil distribution

One specimen found in the John Day Formation in Oregon and was described by Eaton in 1922.
Source (edited): "http://en.wikipedia.org/wiki/Dinaelurus"

Dinocrocuta

Dinocrocuta is an extinct genus of hyena-like feliform carnivores. It lived in Asia, and Africa, during the Miocene epoch. It had very strong jaws that were able to crush bones.

Restoration of *D. giganteus*

It's estimated to weight roughly 400 kg
Source (edited): "http://en.wikipedia.org/wiki/Dinocrocuta"

Dinocyon

Dinocyon is an extinct genus of bear-like hemicyonid carnivore of the Miocene epoch, endemic to Europe living from around 20.3—5.3 Ma, existing for approximately 15 million years.

Taxonomy

Dinocyon It was named by Jourdan (1861). It was assigned to Ursidae by Carroll (1988).

Fossil distribution

- Poysbrunn site, Austria ~20.3—5.3 Ma.

Source (edited): "http://en.wikipedia.org/wiki/Dinocyon"

Dinohippus

Dinohippus (Greek: *Terrible horse*), is an extinct herbivorous mammal belonging to the tribe *Equini*, subfamily *Equinae*, which was endemic to North America from the late Hemphillian stage of the Miocene through the Zanclean stage of the Pliocene (10.3—3.6 mya) and in existence for approximately 6.7 million years.

Taxonomy

Equus was named by Quinn in 1955 and assigned to *Equidae* by Quinn that same year, by B. J. MacFadden (1986) and R. L. Carroll (1988); and to *Equini* by MacFadden.

It was the most common horse in North America and like *Equus*, *Dinohippus* did not have a dished face. It has a distinctive passive "stay apparatus," formed by bones and tendons, to help it conserve energy while standing for long periods. *Dinohippus* is the first horse to show a rudimentary form of this character, providing additional evidence of the close relationship between *Dinohippus* and *Equus*. *Dinohippus* was originally thought to be a monodactyl horse, but a 1981 fossil find in Nebraska shows that some were tridactyl.

Morphology

Three specimens were examined for body mass by M. Mendoza, C. M. Janis, and P. Palmqvist as well as M. T. Alberdi, J. L. Prado, and E. Ortiz-Jaureguizar.

- Specimen 1: 567.7 kg (1251 lbs)
- Specimen 2: 536.5 kg (1182.7 lbs)
- Specimen 3: 224.0 kg (493.8 lbs)

Fossil distribution

Fossil distribution is widespread throughout North America with more than 30 sites from Florida to Alberta, Canada to Central Mexico.

Species

Dinohippus interpolatus

Dinohippus interpolatus (synonomized with *Pliohippus bakeri), D. leardi, D. leidyanus (syn. Pliohippus edensis, Pliohippus osborni*)

- **Taxonomy**: *Hippidium interpolatum* was named by Cope (1893) and said originally to have been placed in. Its type locality is Goodnight Bed Formation. *Hippidium interpolatum* was a grazing and browsing animal.
- **Morphology**: One specimen was examined by M. Mendoza, C. M. Janis, and P. Palmqvist. The body mass was estimated to be 119.7 kg (263.8 lbs).

Mendoza et al. (2006:99) state that D. leidyanus body mass ~200 kg.

Dinohippus mexicanus

Dinohippus mexicanus (synonomized with *Equus mesamexicanus, Hippotigris ocotensis, Protohippus muelleri*),

- **Taxonomy**: was named by Lance in 1950 and recombined as *Asinus mexicanus* by Quinn (1955); it was recombined as *Dinohippus mexicanus* by May and Repenning (1982), MacFadden (1984), MacFadden (1986) and Kelly (1998).
- **Morphology**: One specimen was examined by M. Mendoza, C. M. Janis, and P. Palmqvist. The body mass was estimated to be 259.7 kg (572.5 lbs).

Dinohippus spectans

Dinohippus spectans (synonomized with *Equus mesamexicanus, Hippotigris ocotensis, Protohippus muelleri*),

- **Taxonomy**: was named by Cope in 1880, Zittel (1893), Roger (1896), Matthew (1899), Trouessart (1905), Merriam and Sinclair (1907), Cope and Matthew (1915), Osborn (1918), Merriam and et al. (1925), Hay (1930), Stirton (1940), Quinn (1955), Macdonald (1959) and Shotwell (1963). It was recombined as *Hippidion spectans* by Trouessart (1892), Trouessart (1898) and Hay

(1902); it was recombined as *Protohippus spectans* by Matthew (1909); it was considered a nomen dubium by Macdonald (1992). It was again recombined as *Dinohippus spectans* by Creely and et al. (1982), Kelly and Lander (1988), Hulbert (1989), Kelly (1995) and Kelly (1998).

Sister taxa

Astrohippus, Calippus, *Equus* (synonomized with *Allozebra, Asinus, Onager, Plesippus, Neohippus, Equus semiplicatus, Equus tau*), *Hippidion*, *Onohippidium*, *Pliohippus*, *Protohippus* (syn. *Eoequus*)
Source (edited): "http://en.wikipedia.org/wiki/Dinohippus"

Dissopsalis

Dissopsalis is a genus of extinct predatory mammals of the order Creodonta. The older species, *D. pyroclasticus*, lived in Kenya during the middle Miocene, while the type species, *D. carnifex*, ranged from, Pakistan, India to China during the middle to late Miocene (Barry, 1988).

Dissopsalis is the last known creodont genus. It lived alongside its relative *Hyaenodon weilini*, a member of the very successful genus *Hyaenodon*, during the Miocene in China, and survived to the end of the Miocene, whereas *H. weilini* did not.

Species

- Genus *Dissopsalis*
  - *Dissopsalis carnifex*
  - *Dissopsalis pyroclasticus*

Source (edited): "http://en.wikipedia.org/wiki/Dissopsalis"

Dryopithecus

Dryopithecus was a genus of apes that is known from Eastern Africa into Eurasia the late Miocene period. The first species of *Dryopithecus* was discovered at the site of Saint-Gaudens, Haute-Garonne, France, in 1856. Other dryopithecids have been found in Hungary, Spain, and China.

Like *Sivapithecus*, *Dryopithecus* was suspensory, had a large brain, and a delayed development, but, unlike the former, it had a gracile jaw with thinly enameled molars and suspensory forelimbs; Begun 2004 notes that the similarities and differences between them provides insight into the timing and paleogeography of hominid origins and the phylogenetic divide between Asian and Afro-European great apes.

Description

Dryopithecus was about 60 centimetres (24 in) in body length, and more closely resembled a monkey than a modern ape. The structure of its limbs and wrists show that it walked in a similar way to modern chimpanzees, but that it used the flat of its hands, like a monkey, rather than knuckle-walking, like modern apes. Its face exhibited klinorhynchy, with its face being tilted downwards in profile.

It likely spent most of its life in trees, and was probably a brachiator, similar to modern orangutans and gibbons. Its molars had relatively little enamel, suggesting that it ate soft leaves and fruit, an ideal food for a tree-dwelling animal.

The five-cusp and juvenile fissure pattern of its molar teeth, known as the Y-5 arrangement, is typical of the dryopithecids and of hominoids in general.
Source (edited): "http://en.wikipedia.org/wiki/Dryopithecus"

Dryopithecus fontani

Dryopithecus fontani is an extinct ape that was first discovered in Saint-Gaudens, Haute-Garonne, France, in the 19th century, and was dated to the middle Miocene.

Morphology

Dryopithecus fontani had a dental formula of 2:1:2:3 on both the upper and lower jaw. The incisors of this species were relatively narrow and less spatulate compared to other genera such as Proconsul. The upper molars of this species had a partly developed lingual cingulum. *Dryopithecus fontani* had relatively long upper premolars and relatively broad lower premolars. The lower molars had the Y5 pattern, which is also called the Dryopithecus pattern. The cheek had a thin layer of enamel on them and there were low, rounded cusps. This species had gracile canines and a short premaxilla. The mandible of *Dryopithecus fontani* had an inferior transverse torus but lacked a superior transverse torus. The forelimbs of this species had a reduced olecranon process, a deep humeral trochlea, and the loss of the entepicondylar foramen. *Dryopithecus fontani* had an average body mass of around 35.0 kilograms.

Range

Dryopithecus fontani has been found through the continent of Europe, in locations such as Spain and Hungary.

Diet

Based upon dental morphology *Dryopithecus fontani* was a frugivorous species.
Source (edited): "http://en.wikipedia.org/wiki/Dryopithecus_fontani"

Ectopocynus

Ectopocynus ("strange dog") is an extinct genus of small bone crushing carnivorous mammal similar to a dog of the family Canidae) which inhabited North America during the Oligocene living from 33.3—16.0 Ma and existed for approximately 17.3 million years.

Though a carnivore, dentition suggests this animal was a hypercarnivore or mesocarnivore.

Taxonomy

Ectopocynus was named by Wang (1994). Its type is Ectopocynus simplicidens. It was assigned to Canidae by Wang (1994) and Munthe (1998).

Morphology

Two specimens were examined by Legendre and Roth for body mass. The first specimen was estimated to weigh 16.5 kg (36 lb). The second specimen was estimated to weigh 15.2 kg (34 lb).

Ectopocynus is best described as having simple, robust, and blunt yet massive premolars and reduced or lost cusp on lower molars. In this respect, Ectopocynus has many of the characteristics of *Enhydrocyon*. Remains of *Ectopocynus* are limited to mandibles and teeth only.

Source (edited): "http://en.wikipedia.org/wiki/Ectopocynus"

Ekaltadeta

Ekaltadeta is an extinct genus of giant marsupials related to modern rat-kangaroos.

They are hypothesized to have been either predatory, or omnivorous with a fondness for meat, based on their chewing teeth. This conclusion is based mainly on the size and shape of a large buzz-saw-shaped cheek-tooth, the adult third premolar, which is common to all *Ekaltadeta*. A few species actually did also have long predatory "fangs".

Fossils of the animals include two near complete skulls, and numerous upper and lower jaws.

Source (edited): "http://en.wikipedia.org/wiki/Ekaltadeta"

Ekgmowechashala

Ekgmowechashala (Sioux: "little cat man" or "little fox man") is an extinct genus of primate. With a weight of approximately five pounds, around a foot tall and resembling a lemur, it is the only known North American primate of its time; it lived during the late Oligocene and early Miocene. Its classification remains problematic, and it has been classified as a member of the extinct family Omomyidae (related to tarsiers), the equally extinct Plagiomenidae (related to colugos), and the Adapiformes, the extinct relatives of lemurs and other strepsirrhines. Dentitions from Oregon suggest that it was related to *Rooneyia*, though some scientists saw in them a likeness to *Necrolemur* and *Microchoerus*.

The shape of its teeth, and their likeness to those of raccoons, indicate that it ate soft fruit provided by the warm forests of the Rocky Mountains during the early Miocene.

Fossil evidence of *Ekgmowechashala* was discovered on the Pine Ridge Indian Reservation, an Oglala Sioux Native American reservation in South Dakota. Molars were found in 1981 in the basin of John Day River, and these are in the collection of the Burke Museum of Natural History and Culture; in the summer of 1997 John Zancanella of the Bureau of Land Management found a lower molar in the John Day Fossil Beds National Monument. The material from Oregon and South Dakota is attributed to the only known species, ***Ekgmowechashala philotau***. A single *Ekgmowechashala* tooth from the Toledo Bend Ranch Local Fauna of far eastern Texas may represent a second species.

Source (edited): "http://en.wikipedia.org/wiki/Ekgmowechashala"

Ekorus ekakeran

Ekorus ekakeran is a large extinct mustelid that inhabited late Miocene Kenya.

Standing 60 centimeters tall at the shoulders, its build was not similar to that of modern mustelids. Modern-day weasels have short legs and can only achieve short bursts of speed. The legs of *Ekorus* are built like those of leopards. It appears that before Africa's big cats ruled the savannas, the giant mustelid *Ekorus* chased down its prey, such as the three-toed horse *Eurygnathohippus* and the large pig *Nyanzachoerus*. The reason for this evolution may be related to the Great Rift Valley. Before the rift opened, Kenya was more forested. Eventually as moisture was drawn from the forests by the highlands, the forests opened up into grasslands. The swifter creatures built for open terrain prospered, while the slower forest species perished.

Fossils of large Miocene mustelids have also been discovered in North America and Asia.

Source (edited): "http://en.wikipedia.

org/wiki/Ekorus_ekakeran"

Elephas hysudricus

Elephas hysudricus is a species of extinct elephant which lived in the Siwalik hills of India during the Miocene and Pliocene epochs.

Source (edited): "http://en.wikipedia.org/wiki/Elephas_hysudricus"

Enaliarctos

Enaliarctos is an extinct genus of pinniped.

Prior to the discovery of *Puijila*, the five species in the genus *Enaliarctos* represented the oldest known pinniped fossils, having been recovered from late Oligocene and early Miocene (ca. 24-22 million years ago) strata of California and Oregon. It had a short tail and developed limbs with webbed feet. Unlike modern sea lions, it had a set of slicing carnassials; the presence of slicing (rather than purely piercing teeth in modern fish-eating pinnipeds) suggests that *Enaliarctos* needed to return to shore with prey items in order to masticate and ingest them. Still, *Enaliarctos* had some sea lion-like characteristics such as large eyes, sensitive whiskers, and a specialized inner ear for hearing underwater.

Enaliarctos has been heralded as the ancestor of all known pinnipeds, including the families Otariidae (fur seals and sea lions), Desmatophocidae (extinct seal convergent pinnipeds), Phocidae (true seals), and Odobenidae (walruses). Investigations of the biomechanics of *Enaliarctos* indicate that it used both its forelimbs and hindlimbs during swimming. Modern fur seals and sea lions only use their forelimbs, while true seals primarily use their hindlimbs for aquatic propulsion; lastly, the extant walrus uses both fore- and hindlimbs for swimming. It has been postulated that the condition in *Enaliarctos* is ancestral for all pinnipeds, and that forelimb swimming was lost in true seals while hindlimb swimming was lost in fur seals and sea lions. This is significant because there has been considerable debate as to whether pinnipeds share common ancestry. Interpretation of *Enaliarctos* indicates that all pinnipeds share a common ancestor (which, if it was not *Enaliarctos*, must have been something very similar such as the more newly discovered Puijila darwini.)

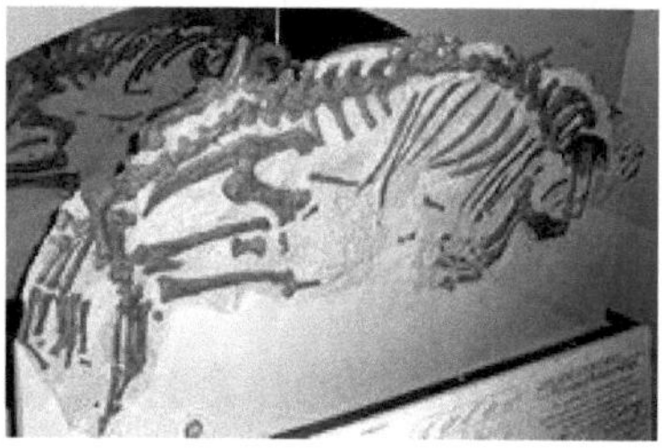

Enaliarctos mealsi

Source (edited): "http://en.wikipedia.org/wiki/Enaliarctos"

Enaliarctos emlongi

Enaliarctos emlongi is an extinct species of pinniped represented by fossils from coastal Oregon. It existed between 13 and 20 million years ago, during the Hemingfordian age of the Oligocene epoch. It was named for renowned fossil collector Douglas Emlong in 1991 by paleontologist Annalisa Berta.

Source (edited): "http://en.wikipedia.org/wiki/Enaliarctos_emlongi"

Enhydritherium terraenovae

Enhydritherium terraenovae is an extinct genus and species of giant otter endemic to North America which lived during the Miocene through Pliocene epochs from ~9.1—4.9 Ma. (AEO), existing for approximately 4.2 million years.

Enhydritherium terraenova ancestral lineage can be traced to Africa and Eurasia but no clear route of immigration can be determined according to Thompson et al.

Taxonomy

Enhydritherium terraenovae was named by Berta and Morgan in 1985 and is the genotype for this animal. Its type locality is the phosphate Palmetto Mine in Florida, which is in a Hemphillian marginal marine sandstone in the Upper Bone Valley Formation of Florida.

Fossil distribution

Fossil specimens were found from in California (3 sites) and Florida (8 sites).

Morpology

Body mass for *Enhydritherium terraenovae* was estimted by Legendre and Roth (1988).

- 16 kg (35 lb)
- 16.4 kg (36 lb)

Source (edited): "http://en.wikipedia.

org/wiki/Enhydritherium_terraenovae"

Enhydrocyon

Enhydrocyon is an extinct genus of bone crushing dog which inhabited North America exclusively during the Oligocene living from around 28.7—22.0 Ma (AEO) and existed for approximately 6.7 million years.

Enhydrocyon's dentition suggests this animal was a hypercarnivore or mesocarnivore.

Taxonomy

Enhydrocyon was named by Cope (1879). Its type is *Enhydrocyon stenocephalus*. It was assigned to Canidae by Cope (1879) and Munthe (1998).

Morphology

Two specimens were examined by Legendre and Roth for body mass. The first specimen was estimated to weigh 22.1 kg (49 lb). The second specimen was estimated to weigh 22.4 kg (49 lb).

Species

- †***Enhydrocyon basilatus*** (23.5—22.7 Ma AEO). One specimen found at the Manville site, Niobrara County, Wyoming. One specimen weighed 70 lbs.
- †***Enhydrocyon crassidens*** (24.4—22.7 Ma AEO). Type locality: Monroe Creek (American Museum of Natural History, Rosebud 12), Shannon County, South Dakota. Weight estimated to be 45-50 lbs. (Legendre & Roth).
- †***Enhydrocyon pahinsintewakpa*** (~24.5—23.3 Ma AEO). Type locality: Sharps SDSM V-5361 aka LACM 1983 (Sharps Formation), Shannon County, South Dakota. Weight was approximately 40-45 lbs. (John Alroy), (Legendre & Roth, S. Legendre and C. Roth. 1988.
- †***Enhydrocyon sectorius*** (*nomen vanum*)
- †***Enhydrocyon stenocephalus*** (25.9—25.7 Ma (AEO). John Day Units G, H, I, and J., Wheeler County, Oregon. Weight 45-50 lbs. (Legendre & Roth).

Source (edited): "http://en.wikipedia.org/wiki/Enhydrocyon"

Enhydrocyon basilatus

Enhydrocyon basilatus is an extinct species of bone crushing hesperocyonid dog that lived during the Late Oligocene to Miocene living from 24.8—20.6 Ma and existed for approximately 4.2 million years.

The dentition of *E. basilatus'* suggests this animal was a hypercarnivore or mesocarnivore.

Taxonomy

Enhydrocyon basilatus was named by Cope (1879). It was recombined as *Hyaenocyon basilatus* by Cope (1879); it was revalidated by Matthew (1907) and Wang (1994). It was assigned to *Hyaenocyon* by Cope (1879); and to *Enhydrocyon* by Cope (1879), Matthew (1907) and Wang (1994).

Morphology

Two specimens were examined by Legendre and Roth for body mass.

- 35.8 kg (79 lb)
- 31.8 kg (70 lb).

Source (edited): "http://en.wikipedia.org/wiki/Enhydrocyon_basilatus"

Entelodont

Entelodonts, sometimes nicknamed **Hell Pigs** or **Terminator Pigs** is an extinct family of pig-like omnivores endemic to forests and plains of North America, Europe, and Asia from the middle Eocene to early Miocene epochs (37.2—16.3 mya), existing for approximately 20.9 million years.

Taxonomy

Entelodont by *Charles R. Knight*

Entelodontidae was named by Richard Lydekker and assigned to *Nonruminantia* by Gregory (1910). Then assigned to *Artiodactyla* by Lucas et al. (1998); and to *Entelodontoidea* by Carroll (1988) and Boisserie et al. (2005). While entelodonts have long been classified as members of the Suina, Spaulding et al. have found them to be closer to whales and hippos than to pigs.

Morphology

Entelodonts are an extinct group of omnivorous mammals, rather pig-like animals, with bulky bodies but short, slender legs, and long muzzles. The largest were the North American *Daeodon shoshonensis* and the Eurasian *Paraentelodon intermedium*, standing up to

2.1 metres (6.9 ft) tall at-shoulder, and had brains the size of an orange.

Body mass

A single specimen was recorded by M. Mendoza, C. M. Janis, and P. Palmqvist for body mass and was estimated to have a weight of 421 kg (930 lb).

Dentition and Anatomy

They had a full set of teeth, including large canines, heavy incisors, and relatively simple, yet powerful, molars. These features suggest an omnivorous diet, similar to that of modern pigs. Like many other artiodactyls, they had cloven hooves, with two toes touching the ground, and the remaining two being vestigial.

The most visible characteristics of the animals, however, would have been the heavy, bony lumps on either side of their heads which are similar to a warthog's. Some of these may be have been attachment points for powerful jaw muscles, but some were larger in males, suggesting that they may also have had a role in competition for mates.

Lifestyle

Entelodonts lived in the forests and plains where they were the apex predators of North America's Early Miocene and Oligocene, consuming carrion and live animals and rounding off their diet with plants and tubers. They would have hunted large animals, like *Eporeodon major* and *Poebrotherium wilsoni*, dispatching them with a blow from their jaws. Some fossil remains of these other animals have been found with the bite marks of entelodonts on them. Like modern day pigs, they were omnivores, eating both meat and plants, but their adaptations show a bias towards live prey and carrion. They most likely were opportunists, mainly eating live animals, but wouldn't sneer at carrion and would eat roots and tubers in times of drought. Some entelodonts even exhibited cacheing behavior, as an *Archaeotherium*'s cache has been discovered, made up of the remains of several early camels.

In popular culture

Entelodonts appear in *Walking with Beasts* and the 2001 remake of *The Lost World*. The creature is also featured in the video game *Wildlife Park 2*, and its colouration is similar to that in the previous films.

Source (edited): "http://en.wikipedia.org/wiki/Entelodont"

Eobalaenoptera

Eobalaenoptera harrisoni is an extinct species of baleen whale. The species was first described in June 2004 by researchers at the Virginia Museum of Natural History.

Artistic impression of a Megalodon pursuing two *Eobalaenoptera*

The researchers identified the species from a partial skeleton found in 1990 in Caroline County, Virginia, the site of a prehistoric ocean, in middle Miocene Calvert Formation rocks. The 11 m (35 ft) skeleton proved to have similar morphological characteristics to a clade of whales consisting of two modern taxonomic families - Balaenopteridae (the rorquals), and Eschrichtiidae (a family with one surviving species, the Gray Whale).

The age of skeleton was estimated to be 14 million years, making the species the oldest known member of the clade by three to five million years. It also considerably narrowed the gap between the earliest known fossil record and estimated time of divergence of this clade from other baleen whales. Molecular studies have put this divergence time at around 25 million years.

The genus name ***Eobalaenoptera*** reflects the similarities between this skeleton and species in the genus *Balaenoptera* such as the Minke Whale; *eo-* is a prefix meaning dawn. The species is named after Carter Harrison, a volunteer worker at the museum.

Source (edited): "http://en.wikipedia.org/wiki/Eobalaenoptera"

Eocardia

Eocardia is an extinct genus of rodent from South America. The 30 centimetres (12 in) long creature was related to guinea pigs and the capybara. It probably resembled a modern marmot.

Source (edited): "http://en.wikipedia.org/wiki/Eocardia"

Eotragus

Eotragus is an early bovid from Europe, Africa, and Asia during the Miocene some 20-18 million years ago. It is related to the modern Nilgai and Four-horned Antelope. It was small and probably lived in woodland environments.

Source (edited): "http://en.wikipedia.org/wiki/Eotragus"

Eporeodon

Eporeodon is an extinct genus of oreodont belonging to the subfamily *Eporeodontidae* during the Oligocene epoch (33.3—20.6 mya) existing for approximately 12.7 million years.

Restoration of *Eporeodon major*

The species of the genus are among the largest members of the family Merycoidontidae. About the size of a cow, its fossils are some of the largest found in the Badlands. It was much larger than *Merycoidodon* and *Miniochoerus*, its contemporaries in the Late Oligocene. On the other hand, *Eporeodon* was a rare oreodont, as its fossils only make up about one percent of all oreodont fossils found in the Badlands.

Taxonomy

Eporeodon was named by Marsh (1875). It was synonymized subjectively with *Eucrotaphus* by Cope (1884). It was assigned to *Merycoidodontidae* by Marsh (1875), Scott (1890), Thorpe (1921), Thorpe (1937), Stevens and Stevens (1996) and Lander (1998). It's synonymized with *Hypselochoerus* and *Pseudodesmatochoerus*.

Morphology

Eporeodon was a tremendous oreodont, growing to be about cow sized. Their skulls were much shorter and squatter than those of *Merycoidodon*, but were longer than those of *Miniochoerus*.

A single specimen was examined by M. Mendoza for body mass and estimated to have a weight of 118.3 kg (260.8 lbs).

Species

E. occidentalis (syn. E. davisi, E. leptacanthus, E. longifrons, E. perbullatus, Eucrotaphus pacificus), E. pygmyus.

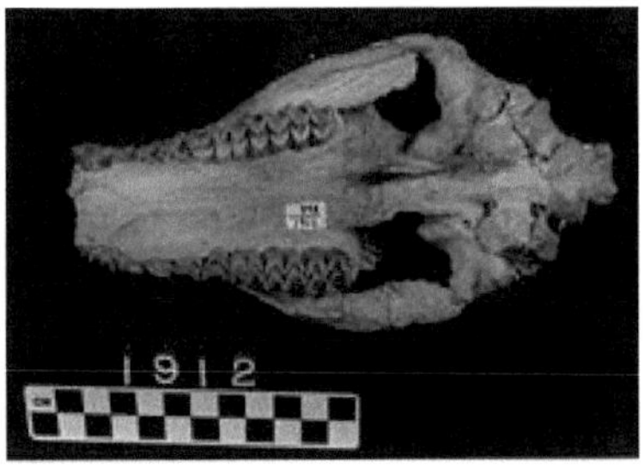

Eporeodon occidentalis skull

Source (edited): "http://en.wikipedia.org/wiki/Eporeodon"

Equinae

Equinae is a subfamily of the family *Equidae*, which lived worldwide (except Indonesia and Australia) from the Hemingfordian stage of the Early Miocene to present (20.6—0 mya) and in existence for approximately 20.6 million years.

Taxonomy

Equinae was named by Steinmann and Döderlein in 1890 and was considered monophyletic by MacFadden (1998).

Subtaxa

The subfamily contains to tribes the Equini and the Hipparionini as well as two unplaced genera *Merychippus* and *Scaphohippus*.

Sister taxa

Anchitheriinae, Hyracotheriinae

Source (edited): "http://en.wikipedia.org/wiki/Equinae"

Equini

Equini is a tribe of herbivorous mammals of the subfamily *Equinae*, family *Equidae*, which lived worldwide (except Australia) from the Hemingfordian stage of the Early Miocene to present (20.6—0 mya) and in existence for approximately 20.6 million years.

Taxonomy

Equini was named by Gray in 1821 and was considered monophyletic by MacFadden (1998).

Subtaxa

Astrohippus, Calippus, Dinohippus, Equus (synonymized with *Allozebra, Asinus, Onager, Plesippus, Neohippus, Equus semiplicatus, Equus tau*), *Hippidion, Onohippidium, Pliohippus, Protohippus* (synomymized with *Eoequus*).

Sister taxa

Hipparionini (synonymized with *Cor-*

mohipparion) Source (edited): "http://en.wikipedia. org/wiki/Equini"

Eubelodon

Eubelodon morrilli is an extinct genus of a North American proboscidean. This gomphothere had a trunk and tusks. It lived during the Miocene Epoch. It may have evolved from Platybelodon.

Fossil distribution

Fossils are restricted to what is now the Great Plains of the United States. Remains were found in the Poison Ivy Quarry, Antelope, Brown County, Nebraska, and Tripp County, South Dakota.

Taxonomy

Eubelodon was named by Erwin Hinckly Barbour in 1914. It was synonymized subjectively with *Trilophodon* by Osborn in 1918 and again by Tobien in 1973 with *Gomphotherium*

It was assigned to Gomphotheriidae by Erwin Barbour in 1914. It was then assigned to Rhynchotheriinae by McKenna and Bell in 1997, Carroll in 1988, Shoshani and Tassy in 1996, Lambert and Shoshani in 1998, and Shoshani and Tassy in 2005.
Source (edited): "http://en.wikipedia. org/wiki/Eubelodon"

Eumegamys

Eumegamys is an extinct genus of dinomyid rodent from the late Miocene and Pliocene of South America.
Source (edited): "http://en.wikipedia. org/wiki/Eumegamys"

Eurhinodelphis

Eurhinodelphis ("well-nosed dolphin") is an extinct genus of Miocene cetacean. Its fossils have been found in France, Belgium, Maryland and California.

Description

Eurhinodelphis longirostris

Eurhinodelphis was around 2 metres (6.6 ft) in length. In most respects, it would have looked like a modern dolphin or porpoise, but its upper jaw was elongated into a sharp tip similar to that of a swordfish. Most likely, *Eurhinodelphis* used it in a similar manner to swordfish, hitting or stabbing prey. It also had long, sharp teeth.

Compared with earlier fossil species, *Eurhinodelphis* had complex ears, suggesting that it already hunted by echolocation like modern whales. Its brain was also asymmetrical, a trait found in modern dolphins, and possibly associated with the complexities of navigating its environment.

Eurhinodelphis was closely related to the orca-sized *Macrodelphinus*.

History

Eurhinodelphis was first described by B. Du Bus in a paper read before the Royal Academy of Sciences of Belgium on 17 December, 1867. O. Abel studied and illustrated the European species in a series of articles published in 1901, 1902 and 1905; subsequently, fossil skulls found in the Calvert Formation in Maryland and Virginia could be attributed to this genus.
Source (edited): "http://en.wikipedia. org/wiki/Eurhinodelphis"

Euroamphicyon

Euroamphicyon (European Amphicyon) is a member of extinct terrestrial carnivores belonging to the family Amphicyonidae, suborder Caniformia, and which inhabited Europe during the Miocene.

Morphology

Eurocyon was one of the larger and heavier Bear-dogs. A specimen was examined by Viranta for body mass and estimated to weigh 138.59 kg (310 lb).
Source (edited): "http://en.wikipedia. org/wiki/Euroamphicyon"

Floridaceras

Floridaceras is an extinct genus of Rhinocerotidae (rhinoceros) of the Miocene epoch, endemic to North

America, living from around ~20.6—16.3 Ma, existing for approximately 4.3 million years.

Taxonomy

Floridaceras was named by Wood (1964). Its type is *Floridaceras whitei.* It was assigned to Rhinocerotidae by Wood (1964) and Carroll (1988); and to Aceratheriinae by Prothero (1998).

Fossil distribution

The only site known is the Thomas Farm Site in Gilchrist County, Florida, ~20.6—16.3 Ma.

Source (edited): "http://en.wikipedia.org/wiki/Floridaceras"

Galadi

Galadi is an extinct genus of predatory bandicoot from Oligo-Miocene deposits of Riversleigh, northwestern Queensland, Australia. It was first named by K.J. Travouillon, Y. Gurovich, R.M.D. Beck and J. Muirhead in 2010 and the type species is *Galadi speciosus.* The genus is represented by three well-preserved skulls and several isolated maxillae and dentaries. Its body mass would have been close to two pounds, making it relatively large for its family. The combination of body size, robustness and short, stout skull indicates that *Galadi* would have been able to take relatively large prey for its size, though the morphology of its molars indicates it may have been omnivorous.

Source (edited): "http://en.wikipedia.org/wiki/Galadi"

Gentilicamelus

Gentilicamelus is an extinct genus of terrestrial herbivore the family Camelidae, endemic to North America from the mid-Oligocene through Early Miocene 30.8—20.6 mya existing for approximately 10.2 million years.

Taxonomy

Gentilicamelus was named by Loomis (1936). It was assigned to Camelidae by Loomis (1936) and Carroll (1988).

Morphology

Four specimens were examined for estimated body mass by M. Mendoza, C. M. Janis, and P. Palmqvist. These specimens were estimated to weigh:

- 52.9 kg (120 lb)
- 77.4 kg (170 lb)
- 64.7 kg (140 lb)
- 43.4 kg (96 lb)

Source (edited): "http://en.wikipedia.org/wiki/Gentilicamelus"

Ginsburgsmilus

Ginsburgsmilus is an extinct genus of carnivorous mammal of the family Barbourofelidae (false saber-tooth cats) endemic to Africa during the early Miocene. There is only one known specimen of *Ginsburgsmilus napakensis*, dated to 20-19 mya.

Taxonomy

Ginsburgsmilus was named by Morales et al. (2001). It was assigned to *Barbourofelidae* by Morlo et al. (2004) and Morlo (2006).

A team led by Jorge Morales described *Ginsburgsmilus* as a new genus in the Barbourofelidae family in 2001 from fossil material previously identified as *Afrosmilus turkanae.*

Source (edited): "http://en.wikipedia.org/wiki/Ginsburgsmilus"

Giraffokeryx

Giraffokeryx was a primitive giraffid that lived in the Miocene age around 27 million years ago in Asia, Europe, and Africa. It had two pairs of elongated ossicones, one pair above its eyes, and the other pair upon the snout. Although it bore a very superficial resemblance to the modern Okapi, it was closely related to the sivatheres *Bramatherium* and *Sivatherium.* Its full scientific name is *Giraffokeryx punnjabiensis.*

Source (edited): "http://en.wikipedia.org/wiki/Giraffokeryx"

Glyptodontidae

Glyptodonts were large, more heavily armored relatives of extinct pampatheres and modern armadillos.

They first evolved during the Miocene in South America, which remained their center of species diversity.

For example, an Early Miocene glyptodont with many primitive features, when compared to other species, *Parapropalaehoplophorus septentrionalis*, was discovered at a now-elevated site in Chile and described in 2007. When the Panama isthmus formed about three million years ago, several species, such as *Glyptotherium texanum*, spread north as part of the Great American Interchange, as did pampatheres and armadillos.

The main feature of glyptodonts was their tortoise-like body armour that was made of bone segments called osteoderms or scutes. Each species of glyptodont had its own unique osteoderm pattern and shell type. With this protection, they were armored like turtles, but unlike most turtles, could not withdraw their heads, but instead had a bony cap on the top of their skull. Even the tail of glyptodonts had a ring of bones for protection. *Doedicurus* even possessed a large mace-like tail that it would have used to defend itself from other *Doedicurus* and predators. Glyptodonts also had size on their side; many such as the type genus, *Glyptodon*, were the size of modern automobiles. Such heavy defenses presupposes a large, effective predator. At the time the glyptodontids evolved, the top predators in the island continent of South America were phorusrhacids, a family of giant flightless carnivorous birds.

The glyptodonts were grazing herbivores. Like many other xenarthrans, they had no incisor or canine teeth, but had a number of cheek teeth that would have been able to grind up tough vegetation, such as grasses. They also had distinctively deep jaws, with large downward bony projections that would have anchored powerful chewing muscles in life.

Glyptodonts went extinct at the end of the last ice age along with a large number of other megafaunal species, including the giant ground sloths and the bizarre *Macrauchenia*. Their much smaller, more lightly armored and flexible relatives, the armadillos, survived.

In physical appearance glyptodonts superficially resembled the much earlier dinosaurian ankylosaurs, and to a lesser degree the recently extinct giant meiolaniid turtles of Australia, examples of the convergent evolution of unrelated lineages into similar forms.

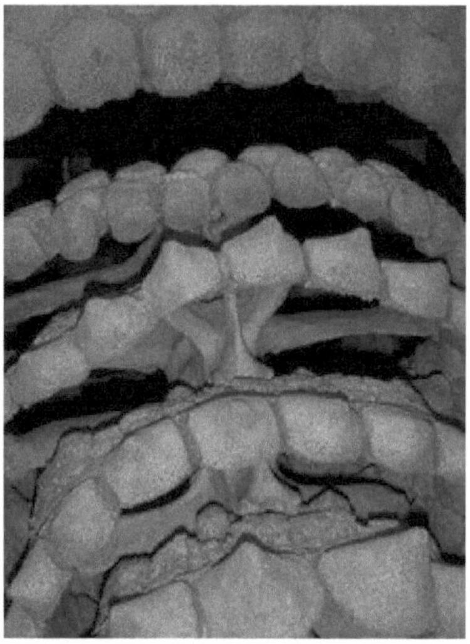

Glyptodont Skeleton: Structural Detail. This armored shell has sections of plated or horned bone between the "shell" (almost like a tortoise shell) and the tail. This area shows where the shell joins the long tail.

Doedicurus clavicaudatus

Propalaeohoplophorus minor remains

Source (edited): "http://en.wikipedia.org/wiki/Glyptodontidae"

Gnathabelodon

Gnathabelodon is an extinct proboscidean (elephant) genus endemic to North America that includes species that lived during the Middle to Late Miocene. They were called "Spoon-billed Mastodons" since their lower jaw was elongated and shaped like a shoehorn or spoon. The genus *Gnathabelodon* consists of two species: **Thorpe's Spoon-billed Mastodon** (*G. thorpei*) and **Buckner's Spoon-billed Mastodon** (*G. buckneri*). The flaring of the tip of their lower jaw was similar to that of the "shovel-tuskers" (Platybelodon and Amebelodon) however *Gnathabelodon* species are distinct in having no lower tusks whilst the "shovel tuskers" have broad, flattened lower tusks. The upper tusks are large and curve outwards and upwards. In respects to dentition and overall body form, it was similar to other species of *Gomphotherium*.

Source (edited): "http://en.wikipedia.

org/wiki/Gnathabelodon"

Gobicyon

Gobicyon is an extinct genus of terrestrial carnivore belonging to the family Canidae ("bear dog") endemic to Central Asia from the Early Miocene subepoch through Late Miocene subepoch 13.6—11.6 Ma, existing for approximately 2 million years.

Gobicyon was named by Colbert in 1839.

Fossil distribution

Tong Xin, Central China.

Sister genera

Alopex, *Caedocyon*, *Canis* (synonymous with *Aenocyon, Lyciscus*), *Cerdocyon*, *Chailicyon*, *Chrysocyon*, *Cuon*, *Cynodesmus*, *Cynotherium*, *Dusicyon*, *Ectopocynus*, *Enhydrocyon* (syn. *Hyaenocyon*), *Eucyon*, *Hesperocyon* (synonymous with *Pseudocynodictis*), *Leptocyon*, *Mesocyon* (synonymous with *Hypotemnodon*), *Neocynodesmus*, *Nyctereutes*, *Osbornodon*, *Otocyon*, *Paraenhydrocyon*, *Philotrox*, *Protocyon*, *Prototocyon*, *Pseudalopex*, *Sunkahetanka*, *Theriodictis*, *Urocyon*, *Vulpes*.

Source (edited): "http://en.wikipedia.org/wiki/Gobicyon"

Gomphotaria

Gomphotaria pugnax ("Pugnacious Wedge-Seal") was a species of very large shellfish-eating dusignathine walrus found along the coast of what is now California, during the late Miocene. It had four tusks, with one pair in the lower and upper jaws, and according to wear on the tusks, *G. pugnax* hammered shellfish open, rather than simply suck them out of their shells as do modern walruses. In terms of the postcranial skeleton, *Gomphotaria* and other dusignathine walruses were built more like sea lions than modern walrus. *Gomphotaria* is a prime example of the extreme diversity that walruses once exhibited.

Source (edited): "http://en.wikipedia.org/wiki/Gomphotaria"

Gomphothere

Gomphothere

Gomphotheriidae is a diverse taxonomic family of extinct elephant-like animals (proboscideans), called **gomphotheres**. They were widespread in North America during the Miocene and Pliocene epochs, 12-1.6 million years ago. Some lived in parts of Eurasia, Beringia and, following the Great American Interchange, South America. From about 5 million years ago onwards, they were slowly replaced by modern elephants, but the last two South American species, *Cuvieronius*, did not finally become extinct until possibly as recently as 9,100 BP, and *Stegomastodon* remains have been dated to 6,060 BP in the Valle del Magdalena, Colombia. Gomphotheres also survived in Mexico and Central America until the end of the Pleistocene.

Gomphothere remains are common at South American Paleo-indian sites. Some examples include the early human settlement at Monte Verde, Chile, dating to approximately 14,000 years ago, and the Valle del Magdalena, Colombia.

An gomphotere remain was found in the El Fin del Mundo site in Sonora, Mexico's clovis location, being the first time such association is found in that part of the continent where gomphoteres were thought to be extinct 30,000 years ago.

Gomphotheres differed from elephants in their tooth structure, particularly the chewing surfaces on the molar teeth. Most had four tusks, and their retracted facial and nasal bones prompt paleontologists to believe that gomphotheres had elephant-like trunks. The early gomphotheres, such as *Phiomia*, had elongated upper and lower jaws, with relatively short tusks. Two lineages appear to have arisen from these ancestors. One, including animals such as *Anancus*, developed the short lower jaw typical of modern elephants, while the others, including *Platybelodon*, developed the lower jaw into an elongated 'shovel', and shortened the upper jaw.

The systematics and phylogeny of gomphotheres are unclear and the group might in fact be paraphyletic. The genus *Gnathabelodon* is often placed in its own family, the Gnathabelodontidae, and *Archaeobelodon*, *Protanancus*, *Amebelodon*, *Platybelodon* and *Serbelodon* are sometimes regrouped in a separate family, the Amebelodontidae. The genera *Anancus*, *Tetralophodon*, *Stegomastodon*, *Paratetralophodon* and *Cuvieronius* are placed by some authors within the gomphotheres, while others consider them as true Elephantidae.

In 1933, *Serbelodon burnhami* (now *Amebelodon burnhami*) was newly discovered and named after the celebrated scout Frederick Russell Burnham.

References

Source (edited): "http://en.wikipedia.org/wiki/Gomphothere"

Gomphotherium

Gomphotherium (pronounced /ˌɡɒm-fəˈθɪəriəm/; "Welded Beast") is an extinct genus of proboscid which evolved in the Early Miocene of North America from 13.650—3.6 Ma, living about 10 million years.

The genus emigrated into Asia, Europe and Africa after a drop in sea level (probably during the Tortonian epoch) allowed them to cross over. It survived into the Pliocene, and its remains have been found in Chile, France, Germany, Austria, Kansas, Tennessee, Pakistan and Kenya.

Description

Gomphotherium angustidens skeleton

Gomphotherium stood around 3 metres (9.8 ft) high, and bore a strong resemblance to a modern elephant. However, it had four tusks; two on the upper jaw and two on the elongated lower jaw. The lower tusks are parallel and shaped like a shovel and were probably used as such. Unlike modern elephants, the upper tusks were covered by a layer of enamel. Compared to elephants, the skull was more elongated and low, indicating that the animal had a short trunk, rather like a tapir's. These animals probably lived in swamps or near lakes, using their tusks to dig or scrape up aquatic vegetation. In comparison to earlier proboscids, *Gomphotherium* had far fewer molars; the remaining ones had high ridges to expand their grinding surface. *Gomphotherium* inhabited dry wooded regions near lakes.

A complete skeleton of *Gomphotherium* has been found at Mühldorf, Germany, in 1971.

Taxonomy

Gomphotherium was named by Burmeister (1837). It is not extant. It was considered paraphyletic by Lambert (2007). It was assigned to Gomphotheriidae by Burmeister (1837), Qiu et al. (1981), Carroll (1988), Lambert and Shoshani (1998), Sach and Heizmann (2001), Sanders and Miller (2002) and Lambert (2007). ***Serridentinus*** was named by Osborn (1923) and synonymized subjectively with Gomphotherium by Tobien (1972), Madden and Storer (1985), Shoshani and Tassy (1996), Lambert (1996), Lambert and Shoshani (1998) and Christiansen (2007).

Source (edited): "http://en.wikipedia.org/wiki/Gomphotherium"

Ground sloth

Ground sloths are a diverse group of extinct sloths, in the mammalian superorder Xenarthra. Their most recent survivors lived in the Antilles, where it has been proposed they may have survived until 1550 CE; however, the youngest AMS radiocarbon date reported is 4190 BP, calibrated to c. 4700 BP for *Megalocnus* of Cuba. They had been extinct on the mainland of North and South America for 10,000 years or more. The term "ground sloth" is used as a reference for all extinct sloths because of the large size of the earliest forms discovered, as opposed to the extant "tree sloths." However, this is a historical convention and does not imply that all extinct sloths were strictly terrestrial in nature.

The bulk of ground sloth evolution took place during the mid to late Tertiary of South America while the continent was isolated. At their earliest appearance in the fossil record, the ground sloths are already distinct at the family level. The presence of intervening islands between the American continents in the Miocene allowed a dispersal of forms into North America. A number of mid- to small-sized forms are believed to have previously dispersed to the Antilles islands either by making short swims or using land bridges that are now submerged. Ground sloths were a hardy group as evidenced by their diverse numbers and dispersals into remote areas given the finding of their remains in Patagonia (Cueva del Milodón) and parts of Alaska.

Sloths, and xenarthrans as a whole, represent one of the more successful South American groups during the Great American Biotic Interchange. During the interchange, many more taxa moved from North America into South America than in the other direction. At least five genera of ground sloths have been identified in North American fossils; these are examples of successful immigration to the north.

Families

Paleontologists divide more than 80 genera of ground sloths in multiple families.

Megalonychidae

Reconstruction based on *Megalonyx jeffersonii*, Iowa Museum of Natural History, University of Iowa.

The megalonychid ground sloths first appeared in the early Oligocene, about 35 million years ago, in southern Argentina (Patagonia). Megalonychids first reached North America by island-hopping, prior to the formation of the Isthmus of Panama. Some lineages of megalonychids increased in size as time progressed. The first species of these were small and may have been partly tree-dwelling, whereas the Pliocene (about 5 to 2 million years ago) species were already approximately half the size of the huge Late Pleistocene *Megalonyx jeffersonii* from the last ice age. Some West Indian island species were as small as a large cat; their dwarf condition typified both tropical adaptation and their restricted island environment. This small size also enabled them a degree of aboreality.

Megalonyx, which means "giant claw" is a widespread North American genus, lived past the close of the last (Wisconsin) glaciation, when so many large mammals died out. Remains have been found as far north as Alaska and the Yukon. Ongoing excavations at Tarkio Valley in southwest Iowa may reveal something of the familial life of *Megalonyx*. An adult was found in direct association with two juveniles of different ages, suggesting that adults cared for young of different generations.

The earliest known North American megalonychid, *Pliometanastes protistus*, lived in Florida and the southern U.S. about 9 million years ago, and is believed to have been the predecessor of *Megalonyx*. Several species of *Megalonyx* have been named; in fact it has been stated that "nearly every good specimen has been described as a different species". A broader perspective on the group, accounting for age, sex, individual and geographic differences, indicates that only three species are valid (*M. leptostomus*, *M. wheatleyi*, and *M. jeffersonii*) in the late Pliocene and Pleistocene of North America. Although work by McDonald lists five species. Jefferson's ground sloth has a special place in modern paleontology, for Thomas Jefferson's letter on *Megalonyx*, read before the American Philosophical Society of Philadelphia, in August 1796, marked the beginning of vertebrate paleontology in North America. When Lewis and Clark set out, Jefferson instructed Meriwether Lewis to keep an eye out for ground sloths. He was hoping they would find some living in the Western range. *Megalonyx jeffersonii* was appropriately named after Thomas Jefferson.

Megatheriidae

Fossil *Eremotherium* skeleton, National Museum of Natural History, Washington, DC.

The megatheriid ground sloths are relatives of the megalonychids, with which they form the infraorder Megatheria. Megatheriids appeared later in the Oligocene, some 30 million years ago, also in South America. The group includes the heavily-built *Megatherium* (given its name 'great beast' by Georges Cuvier) and *Eremotherium*. The skeletal structure of these ground sloths indicates that the animals were massive. Their thick bones and even thicker joints (especially those on the hind legs) gave their appendages tremendous power that, combined with their size and fearsome claws, provided a formidable defense against predators.

The earliest megatheriid in North America was *Eremotherium eomigrans* which arrived 2.2 million years ago, after crossing the recently formed Panamanian land bridge. With more than five tons in weight, 6 meters in length, and able to reach as high as 17 feet (5.2 m), it was taller than an African Bush Elephant bull. Unlike relatives, this species retained a plesiomorphic extra claw. While other species of *Eremotherium* had four fingers with only two or three claws, *E. eomigrans* had five fingers, four of them with claws up to nearly a foot long.

Nothrotheriidae

Recently recognized, members of Nothrotheriidae are often associated with those of the Megatheriidae, and together the two form the superfamily Megatheroidea. The most prominent members of the group are the South American genus *Thalassocnus*, known for being aquatic, and *Nothrotheriops* from North America.

The last ground sloths in North America belonging to *Nothrotheriops* died so recently that their petrified dung (coprolites) has remained undisturbed in some caves, as if it were just recently deposited. One of the skeletons, found in a lava tube (cave) at Aden Crater, adjacent to Kilbourne Hole, New Mexico, still had skin and hair preserved, and is now at the Yale Peabody Museum. The American Museum of Natural History in New York City has a sample of dung with a note attached to it that reads "deposited by Theodore Roosevelt". The largest samples of *Nothrotheriops* dung can be found in the collections of the Smithsonian Museum.

Mylodontidae

Another specimen of *Paramylodon harlani*. Texas Memorial Museum, University of Texas at Austin.

The mylodontid ground sloths together with their relatives the scelidotheriids and the orophodontids form the Mylodonta, the second radiation of ground sloths. The discovery of their fossils in caverns associated with human occupation lead some early researchers to theorize that the early humans built corrals when they could procure a young ground sloth, to raise the animal to butchering size. However, radiocarbon dates do not support simultaneous occupation of the site by humans and sloths. Subfossil remains like coproliths, fur and skin have been discovered in some quantities.

Orophodontidae

The orophodontid ground sloths constitute a rather small but quite distinct group. They are classified as a distinct mylodontan superfamily **Orophodontoidea**, the sister taxon to the Mylodontoidea.

In cryptozoology

Cryptozoologists often identify the *mapinguari*, a mythical forest creature of the upper Amazon basin, with a surviving tropical ground sloth or folk memory of these animals.

Source (edited): "http://en.wikipedia.org/wiki/Ground_sloth"

Guiomys

Guiomys is an extinct genus of cavioid rodent which existed in west central Patagonia of Argentina, during the middle Miocene. *Guiomys* is known from mandibular and maxillary fragments with molars, and isolated cheek teeth. It was first named by María E. Pérez in 2010 and the type species is *Guiomys unica*.

Source (edited): "http://en.wikipedia.org/wiki/Guiomys"

Gulfoceras

Gulfoceras was a genus of rhinoceros endemic to North America from the Miocene, living from 23.03—20.4 mya existing for approximately 2.63 million years.

Taxonomy

Gulfoceras was named by Albright (1999). Its type is *Gulfoceras westfalli*. It was assigned to Rhinocerotidae by Albright (1999).

Source (edited): "http://en.wikipedia.org/wiki/Gulfoceras"

Haplocyon

Haplocyon is an extinct genus of terrestrial carnivores belonging to the suborder Caniformia, family Amphicyonidae ("bear dog") named by Schlosser in 1901.

Sister genera

Afrocyon, Amphicyanis, Arctamphicyon, Brachycyon, Cynodictis, Drassonax, Goupilictis, Haplocyonoides, Haplocyonopsis, Harpagophagus, Ictiocyon, Pachycynodon, Paradaphoenus, Protemnocyon, Pseudamphicyon, Pseudarctos, Pseudocyonopsis, Sarcocyon, Symplectocyon, Vishnucyon

Source (edited): "http://en.wikipedia.org/wiki/Haplocyon"

Haplocyonoides

Haplocyonoides is an extinct family of terrestrial carnivores belonging to the suborder Caniformia, family Amphicyonidae ("bear dog"), and which inhabited Europe from the Early Miocene subepoch (20 Mya)—(16.9 Mya). *Haplocyonoides* existed for approximately 3.1 million years.

Haplocyonoides was named by Hürzeler (1940) and was assigned to Amphicyonidae by Carroll (1988).

Fossil distribution

One site in Les Beilleaux, France.

Sister genera

Afrocyon, Amphicyanis, Arctamphicyon, Brachycyon, Cynodictis, Drassonax, Goupilictis, Haplocyon, Haplocyonopsis, Harpagophagus, Ictiocyon, Pachycynodon, Paradaphoenus, Protemnocyon, Pseudamphicyon, Pseudarctos, Pseudocyonopsis, Sarcocyon, Symplectocyon, Vishnucyon

Source (edited): "http://en.wikipedia.org/wiki/Haplocyonoides"

Hegetotheriidae

Hegetotheriidae is an extinct family of notoungulate mammals known from the Eocene through the Pleistocene of South America

Source (edited): "http://en.wikipedia.org/wiki/Hegetotheriidae"

Helladotherium

Helladotherium is an extinct genus of Sivatherine Giraffid from Europe, Africa, and Asia during the Miocene. The most complete skeleton is that of a female, based on a comparison with an intact female *Sivatherium giganteum* skull.

Source (edited): "http://en.wikipedia.org/wiki/Helladotherium"

Hemiauchenia

Hemiauchenia is a genus of lamine camelids that evolved in North America in the Miocene period approximately 10 million years ago. This genus diversified and moved to South America in the early Pleistocene as part of the Great American Interchange, giving rise to modern lamines. The genus went extinct at the end of the Pleistocene.

Remains of these species have been found in assorted locations around North America including: Florida, Texas, Kansas, Nebraska, Arizona, Mexico, California, Oklahoma, New Mexico, Colorado and Washington. The "large-headed llama", *H. macrocephala*, was widely distributed in N. and Central America, with *H. vera* being known from the western U. S. and northern Mexico. *H. minima* has been found in Florida, and *H. guanajuatensis* in Mexico.

Distinguishing characteristics of members of *Hemiauchenia*

vera:

- Relatively low-crowned teeth
- Large caniniform upper P1
- Retention of lower P3

blancoensis:

- Shorter mandibular diastema than *macrocephala* but shorter than *vera*
- Canniform upper P1
- Absent P2
- Upper P3 present or absent
- Lower crowned molars

macrocephala:

- Long, robust limbs
- Large skeletal size
- Presence of a deciduous upper P2
- Fully molariform deciduous P2
- High-crowned molars
- Thick layer of cementum on the teeth
- Broad mandibular symphasis with incisors in a vertical fashion

minima:

- Despite being the earliest recognized species, general distinguishing characteristics for *H. minima* are little known.

There are also a few lesser known species such as: *H. paradoxa*, *H. seymourensis*, *H. edensis* and *H. guanajuatensis*. According to which source is consulted, these may or may not be considered legitimate taxa.

Source (edited): "http://en.wikipedia.org/wiki/Hemiauchenia"

Hemicyon

Hemicyon the so-called "dog-bear," literally "Half Dog" (Greek:ἡμικυων/ "hemi-kuôn"), is an extinct genus of the family *Hemicyonidae*, endemic to Europe and Asia during the Miocene epoch (23.03 to 5.33 million years ago), existing for approximately 16.9 to 11.1 mya.

Morphology

Hemicyon was about 1.5 metres (5 ft) long, and 70 centimetres (28 in) tall, with somewhat tiger-like proportions and dog-like teeth. *Hemicyon* is widely accepted to have been hypercarnivorous and highly predaceous. Unlike modern bears, Hemicyon walked on its toes, that is, it was not plantigrade, but digitigrade, with long metapodials. This suggests that Hemicyonidae must have been an active hunter and a good runner, and presumably hunted on the plains, possibly in packs. This also distinguishes it from the true bears, which are plantigrade, and gives it a greater affinity to dogs, which are digitigrade.

Body mass

Two specimens were examined by Legendre and Roth for body mass. The first specimen was estimated to have a weight of 67.2 kg (150 lb). The second specimen was estimated to have a weight of 534.5 kg (1,200 lb).

Fossil sites

Species of *Hemicyon* are recognized in the medial Hemingfordian (ca. 18 million years ago) Thomas Farm local fauna. This record of Hemicyon represents the earliest occurrence of this genus in North America. The Thomas Farm Cynelos Hemicyon is very closely related to European representatives from the early Burdigalian (ca. 20 million years ago). Other *Hemicyon* fossil finds include: *H. teilhardi* from the Middle Miocene Hujialian Formation in the Linxia Basing, Gansu Province, China; an Early Miocene *Hemicyon* from Lanzhu Basin Shan Mong fauna, China; *Hemicyon* of the Mid-Miocene Dongxiang and Laogou Formation in Linxia Basin on the northeast margin of the Tibetan Plateau; Mid-Miocene *H. sansaniensis* of Arroyo del Val area, Villafeliche, Zaragoza Province, Spain; Mid-Miocene *H. mayorali* of Tarazona de Aragon, Spain; a *Hemicyon* found at the Somosaguas site, Pozuelo de Alarcón, Madrid; *H. sansaniensis* from the Mid-Miocene Anchitherium fauna of İnönü locality, NW Ankara; a Lower Miocene (Eggenburgian-Ottnangian: ca. 20 Ma) *H stehlini* found in the coal bed of Banovići Basin, Bosnia; an Early Hemingfordian *Hemicyonid* (ca. 16 Ma) from the Pollack Farm Fauna, Delaware; and an excellent *Hemicyon* specimen from the Miocene Sante Fe Formation, New Mexico.

Hemicyonidae fossil evidence may also have been found in the Gaillard Cut Local Fauna, former Canal Zone, Republic of Panama, but it is indeterminant with Amphicyonidae. Despite its close proximity to South America, given the geological context, the Gaillard Cut Local Fauna are distinctly North American fauna and the age of this assemblage is dated between 19.5 and 14 Ma.

Source (edited): "http://en.wikipedia.org/wiki/Hemicyon"

Hemicyonidae

Hemicyonidae is an extinct family of so-called "dog-bears", literally "Half Dog" (Greek: ἡμικυων "hemi-kuôn"), bear-like carnivoran living in Europe, North America, Africa, and Asia during the Oligocene through Miocene epochs 33.9—5.3 Ma, existing for approximately 28.6 million years.

Morphology

They were about 1.50 m (5 ft) long, 70 centimetres (28 inches) tall, with somewhat tiger-like proportions and dog-like teeth, weighing as much as 69 kg (150 lb).

Hemicyonids are widely accepted to have been hypercarnivorous and highly predaceous. They walked on their toes, ie., they were digitigrade, with long metapodials. In these respects they were similar to dogs, and unlike modern bears, which are plantigrade with short metapodials. This suggests that hemicyonids must have been active hunters and good runners, and presumably hunted on the plains, possibly in packs. Note that the hemicyonids are easily confused with the related and often similar-looking "bear-dogs" of the family Amphicyonidae.

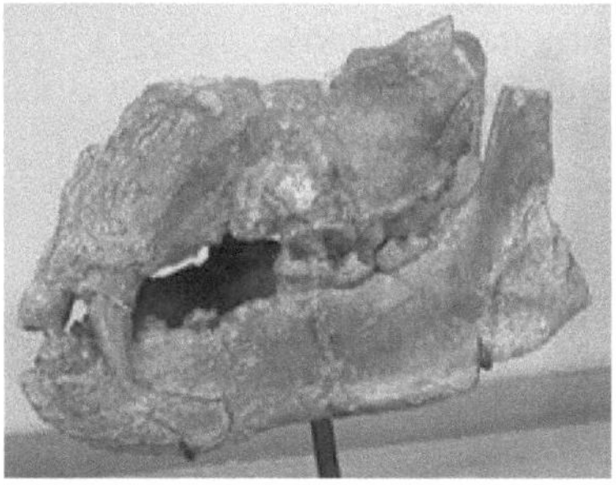

Plithocyon armagnacensis skull

Taxonomy

Classification of the hemicyonids has become complicated. Some classifications subdivide the Hemicyonidae into two subfamilies: Phoberocyoninae (including *Phoberocyon* and *Plithocyon*) and Hemicyoninae (including *Zaragocyon*, *Hemicyon*, and *Dinocyon*). The earliest known occurrence of the hemicyonids is in Europe from 22.8–20 Ma of Spain: *Phoberocyon hispanicus* from Loranca and *Zaragocyon daamsi* from Cetina de Aragon.

Genus identification

Hemicyonids lived continuously in Europe with the oldest known being Cephalogale existing from ~33.9—20 Ma. In North America specimens have been uncovered which date from ~21.6—21.2 Ma. *Phoberocyon johnhenryi* is yet another old specimen from the large Thomas Farm Site, Florida, (~19–18 Ma). This species of the genus *Phoberocyon* is confirmed by diagnostic features of the lower dentition: posterior accessory cusps in lower premolars and paraconid in m2, and it is a more derived species than *Phoberocyon hispanicus* in that it is larger and has diastemata between the premolars.

All the American hemicyonids are more advanced in character, and are assigned to genus *Plithocyon*. They are known from the Barstovian of California, New Mexico, Nebraska, Wyoming and the Love Bone Bed Local Fauna in Florida. In Asia, the earliest hemicyonid is *Phoberocyon youngi* from Shanwang, Shandong (16 Ma) and is very close to the North American *Phoberocyon johnhenry*. The latest occurrence in China is from the Tung-Gur Formation, Nei-Mongol (13 Ma). Dispersal probably occurred from Europe to North America prior to 19 Ma and the American *Plithocyon* is probably a descendant of the *Phoberocyon*.

The genus *Agriotherium* is sometimes considered a member of Hemicyonidae due to the presence of many similar hemicyonine characters (unelongated molars, premasseteric fossa, etc.) and this strongly substantiates its closer affinity with hemicyonines rather than with ursines. *Agriotherium* has its longest history in Asia. The most primitive and earliest form is from a *Hipparion* fauna in Hezheng, Gansu, 9 Ma. All other forms of *Agriotherium* occurred in the Pliocene (5.3 to 2 Ma) in Eurasia and Africa, leaving a hiatus in the record between 9 and 5.3 Ma. In North America, *Agriotherium* appeared suddenly around 7 Ma and probably emigrated from Eurasia. *Agriotherium* fossils have been found at Coffee Ranch, Texas (6.6 Ma); Guymon Local Fauna, Oklahoma; Mount Eden Local Fauna, California; and Old Cabin Quarry, Quiburis Formation (late Hemphillian), Arizona. The occurrence of *Agriotherium* in North and Central America is thus limited to the later half of the Hemphillian (7–5 Ma).

Hemicyonidae fossil evidence may also have been found in the Gaillard Cut Local Fauna, former Panama Canal Zone, Republic of Panama, but it is indeterminant with Amphicyonidae. Despite its close proximity to South America, given the geological context, the Gaillard Cut Local Fauna are distinctly North American fauna and the age of this assemblage is dated between 19.5 and 14 Ma.

Source (edited): "http://en.wikipedia.org/wiki/Hemicyonidae"

Hemicyoninae

Hemicyoninae is an extinct subfamily of bear-like mammals of the family Ursidae endemic to North America and Europe during Oligocene through Miocene, living from ~33.9—7.2 Ma, existing for approximately 26.7 million years.

Hemicyoninae includes the genera Cephalogale, Phoberocyon, and Plithocyon.

Taxonomy

Hemicyoninae was named by Frick (1926) and was assigned to Ursidae by Qiu et al. (1991) and Hunt (1998).

Source (edited): "http://en.wikipedia.org/wiki/Hemicyoninae"

Hemimastodon

Hemimastodon ("half mastodont") is an extinct genus of proboscidea from the Late Miocene deposits of the Dera Bugti Beds in Pakistan.

Its phylogenetic affinity within the proboscidea is uncertain.

Source (edited): "http://en.wikipedia.org/wiki/Hemimastodon"

Herpestides

Herpestides is an extinct genus of terrestrial carnivore the family *Hyaenidae,* endemic to North Africa and Southern Europe during the Early Miocene subepoch (22.4—20 mya) existing for approximately 2.4 million years.

Taxonomy

Herpestides was named by de Beaumont (1967). It is not extant. It was assigned to *Aeluroidea* by Hunt (1989); and to *Viverridae* by Flynn (1998) and Morlo et al. (2007).

Species

H. aegypticus, H. aequatorialis, H. antiquus, H. compactus

Sister genera

Africanictis, Anictis, Asiavorator, Haplogale, Mioprionodon, Moghradictis, Nandinia, Palaeoprionodon, Proailurus, Shandgolictis, Stenogale, Stenoplesictis.

Source (edited): "http://en.wikipedia.org/wiki/Herpestides"

Herpetotherium

Herpetotherium is an extinct genus of pygmy opossum. Species were arboreal omnivores, much like today's opossums.

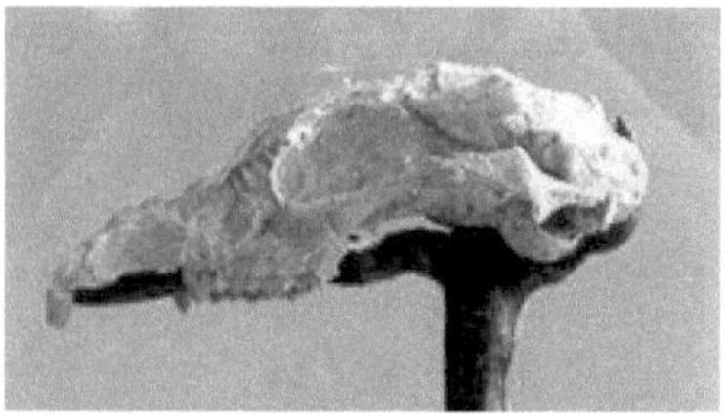
Peratherium fugax

Restoration

Peratherium blainvillei lower jaw at the Museum für Naturkunde, Berlin

Fossils have been found in California, Oregon, Texas, Florida, Montana, Wyoming, Colorado, North and South Dakota, Nebraska, and Saskatchewan. The oldest species, *H. knighti*, is dated to around 50.3mya, and the most recent, an unnamed species, may be as recent as 16.3mya.

Species

- *H. fugax* (syn. *Didelphys pygmaea*, *H. scalare*, *H. tricuspis*, *Peratherium fugax*) (type)
- *H. huntii*
- *H. knighti* (syn. *Centracodon delicatus*, *Entomacodon minutus*, *Peratherium morrisi*)
- *H. merriami*
- *H. valens* (syn. *Peratherium donahoei*)
- *H. youngi* (syn. *Peratherium spindleri*)

Source (edited): "http://en.wikipedia.org/wiki/Herpetotherium"

Hesperocamelus

Hesperocamelus is an extinct genus of terrestrial herbivore the family Camelidae, endemic to North America from the Pliocene through Pleistocene 23.03—5.33 mya existing for approximately 17.7 million years.

Taxonomy

Hesperocamelus was named by Macdonald (1949). It was assigned to Camelidae by Macdonald (1949) and Carroll (1988).

Fossil distribution

Fossil distribution is restricted to Nevada and California.

Source (edited): "http://en.wikipedia.org/wiki/Hesperocamelus"

Heteroprox

Heteroprox is an extinct genus of cervid from the Miocene of Europe.

Source (edited): "http://en.wikipedia.org/wiki/Heteroprox"

Hipparion

Painting of *Hipparion.*

Skull of *Hipparion gracile.*

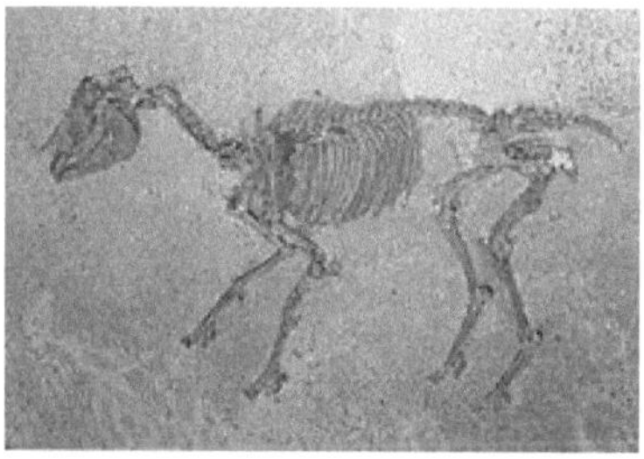
Skeleton of *Hipparion primigenius.*

Hipparion (Greek, "pony") is an extinct genus of horse living in North America,

Asia, Europe, and Africa during the Miocene through Pleistocene ~23 Mya—781,000 years ago, existing for 22.219 million years (genus including species). *Hipparion sp.* (~12.7 Ma—781,000) existed for approximately 11.919 million years.

Its habitat or biome would be that of non-forested, grassy plains, shortgrass prairie or steppes.

Taxonomy

Hipparion was named by de Christol (1832) with it assigned the type European *H. prostylum*. It was assigned to Equidae by de Christol (1832), Thurmond and Jones (1981) and Carroll (1988); and to Hipparionini by MacFadden (1998).

Morphology

Hipparion resembled the modern horse, but still had three vestigal outer toes (in addition to its hoof). These did not touch the ground. *Hipparion* was about 1.4 metres (13.3 h) tall at the shoulder.

Body mass

Three specimens were examined by Legendre and Roth for body mass.

- Specimen 1 was estimated to weigh: 118.9 kg (260 lb)
- Specimen 2 was estimated to weigh: 69.4 kg (150 lb)
- Specimen 3 was estimated to weigh: 62.8 kg (140 lb)

Fossil distribution

- **H. concudense** was named by Pirlot (1956). Fossil distribution: Casa del Acero site, Murcia, Spain and Masia de La Roma site, Aragon, Spain. It was endemic to Southwestern Europe ~11.6—5.3 Mya.
- **H. crassum** was named by Gervais (1859). Fossil distribution: Odessa, Ukraine, Dorkovo, Ukraine. It was endemic to Southeastern Europe ~4.9 Mya—11,000 years ago.
- **H. dietrichi** was named by Wehrli (1941). Fossil distribution: Macedonia, Greece and Nikiti, Greece. It was endemic to Southeastern Europe ~9 —5.3 Mya.
- **H. fissurae** was named by Crusafont and Sondaar (1971). Fossil distribution: Soria, Spain. It was endemic to Southwestern Europe ~5.3 —3.4 Mya.
- **H. forcei** was named by Richey (1948). Fossil distribution: Contra Costa, Santa Cruz, Los Angeles counties. It was endemic to western North America ~23—5.3 Mya.
- **H. gromovae** was named by Villalta and Crusafont (1957). Fossil distribution: Murcia, Spain. It was endemic to Southwestern Europe. ~9—5.3 Mya.
- **H. laromae** was named by Pesquero et al. (2006). Fossil distribution: Teruel, Spain. It was endemic to Southwestern Europe. ~9.7—8.7 Mya.
- **H. longipes** was named by Gromova (1952). Fossil distribution: Ankara, Turkey. It was endemic to Eurasia. ~5.3—3.4 Mya.
- **H. macedonicum** was named by Koufos (1984). Fossil distribution: Greece. It was endemic to Southwestern Europe. ~8.7—7.8 Mya.
- **H. matthewi** was named by Abel (1926). Fossil distribution: Ankara, Turkey. It was endemic to Asia Minor. ~11.6—5.3 Mya.
- **H. mediterraneum** was named by Roth and Wagner (1855). Fossil distribution: Iraq, Turkey, Greece. It was endemic to Asia Minor, Greece ~9—5.3 Mya.
- **H. molayanense** was named by Zouhri (1992). Fossil distribution: Afghanistan It was endemic to Southwest Asia ~9—5.3 Mya.
- **H. periafricanum** was named by de Christol (1832). Fossil distribution: Aragon, Spain. It was endemic to Southwestern Europe ~9—5.3 Mya.
- **H. rocinantis** was named by Hernández Pacheco (1921). Fossil distribution: Spain. It was endemic to Southwestern Europe ~5.3—1.8 Mya.
- **H. sellardsi** was named by Matthew and Stirton (1930).
- **H. shirleyae** was named by MacFadden (1984).
- **H. tehonense** was named by Stirton (1940). Fossil distribution: North America, widespread from Florida to California. ~23.3—5.3 Mya.

Source (edited): "http://en.wikipedia.org/wiki/Hipparion"

Hippopotamus gorgops

Hippopotamus gorgops ("Gorgon-eyed river horse") is an extinct species of hippopotamus. It first appeared in Africa during the late Miocene, and eventually migrated into Europe during the early Pliocene (where its fossils were first discovered). It became extinct prior to the Ice Age.

With a length of 4.3 metres (14 ft) and a shoulder height of 2.1 metres (6.9 ft) *H. gorgops* was much larger than its living relative, *H. amphibius*. Another feature setting it apart from *H. amphibius* were its eyes. Modern hippos have eyes placed high on the skull, but *H. gorgops* took things a step further and had what could be described as short eye stalks, making it even easier for the creature to see its surroundings while (almost) fully under water.

Source (edited): "http://en.wikipedia.org/wiki/Hippopotamus_gorgops"

Hippotherium

Hippotherium is an extinct genus of horse endemic to North America, to Asia, Europe, and Africa during the

Miocene through Pliocene ~13.65—3.3 Mya, existing for 10.35 million years.

Taxonomy

Hippotherium was named by Kaup in 1832 and assigned to Equidae by Kaup 1833. It was named as a subgenus of Equus; type *European E. gracilis*. Its type is ***Equus primigenius***. It was synonymized subjectively with Hipparion by MacFadden (1998) and Pesquero et al. (2006). It was assigned to the tribe Hipparionini by Alroy in 2002 and 2008.

It was synonymized subjectively with Cormohipparion occidentale by MacFadden (1984) and was considered a nomen dubium by Macdonald (1992).

Morphology

Body mass

A single specimen was examined by M. Mendoza, C. M. Janis, and P. Palmqvist for body mass.

- Specimen: 61.5 kg (140 lb)

Fossil distribution

- Rattlesnake site, aka (SDSM V-832) Grant County, Oregon estimated age: ~7.8—7.7 Mya.

Species

H. emsliei

M. elatus is exclusive to the Southeastern U.S. It was named *Cormohipparion emsliei* by Hulbert, Jr. in 1988 and in 2006 by Hulbert and Whitmore in 2007. It became *Hippotherium emsliei* in 2001 by Alroy (J. Alroy).

Body mass

A single specimen was examined by M. Mendoza, C. M. Janis, and P. Palmqvist for body mass. (Mendoza et al)

- Specimen: 48.7 kg (110 lb)

Fossil distribution

- Mauvilla site aka Ecor Rouge Sand, Mobile County, Alabama estimated age: ~10 Mya.
- Haile XVA aka Haile 15A, Alachua County, Florida, estimated age: ~3.7—3.6 Mya.
- Agrico Pierce Mine, Polk County, Florida, estimated age: ~10 Mya.
- Moss Acres Racetrack, Marion County, Florida estimated age: ~10.3—4.9 Mya.

H. ingenuum

H. ingenuum is exclusive to Central America and North America. It was named by Leidy (1885). Its type locality is Mixson's Bone Bed, which is in a Hemphillian terrestrial horizon in the Alachua Formation of Florida.

It was recombined as Hipparion ingenuum by Spencer (1895), Gidley (1907), Osborn (1918), Hay (1930) and Simpson (1930); it was synonymized subjectively with *Hippotherium gratum* by Lucas (1896). In 1909 it was recombined as Neohipparion ingenuum by Matthew, recombined as Nannippus ingenuus by Stirton (1940), Quinn (1955), Forsten (1975) and MacFadden (1984); it was recombined as Cormohipparion ingenuum by Hulbert (1988); it was considered a nomen dubium by Macdonald (1992).

Body mass A single specimen was examined by M. Mendoza, C. M. Janis, and P. Palmqvist for body mass. (Mendoza et al).

- Specimen: 55.3 kg (120 lb)

Fossil distribution

- Las Culebras, Rancho Lobo, and Tepusuna, Honduras, Gracias Formation, estimated age: ~11.6—8.6 Mya.
- TMM 31204 aka John Dye Farm, Goliad County, Texas, estimated age: ~11.6—8.6 Mya.
- Occidental Mine, Statenville Formation, Hamilton County, Florida estimated age: ~11.7 Mya.
- Hog Creek Sarasota County, Florida estimated age: ~11.5—10.1 Mya.

H. plicatile

Hippotherium plicatile is exclusive to the state of Florida in North America. It was named by Leidy (1887). Its type locality is Mixson's Bone Bed, which is in a Hemphillian terrestrial horizon in the Alachua Formation of Florida.

It was recombined as Hipparion plicatile by Spencer (1895), Roger (1896), Trouessart (1898), Hay (1902), Gidley (1907), Sellards (1913), Osborn (1918), Matthew (1924), Simpson (1930), Stirton (1940) and Webb and Perrigo (1984); it was recombined as Neohipparion plicatile by Trouessart (1904), Matthew (1909) and Merriam (1913); it was recombined as *Cormohipparion plicatile* by Hulbert (1988); it was considered a nomen dubium by Macdonald (1992).

Body mass

A single specimen was examined by M. Mendoza, C. M. Janis, and P. Palmqvist for body mass. (Mendoza et al).

- Specimen: 80.1 kg (180 lb).

Fossil distribution

- Hog Creek Sarasota County, Florida estimated age: ~11.5—10.1 Mya.
- Haile 6A, Haile 5B, Gainesville Creek, Pareners Branch, and Coffrin Creek sites, Alachua County, Florida, estimated age: ~11.5—9.3 Mya.
- Withlacoochee River Site 4X, Citrus County, Florida estimated age: ~10.3—8.6 Mya.

H. primigenius

E. primigenius is exclusive to Europe and Africa. Its range extended from Northern Europe to South Africa. It was named by von Meyer. It is the type species of *Hippotherium*. It was recombined as *Hipparion primigenium* by Pesquero et al. (2006); it was recombined as *Hippotherium primigenium* by Kaup (1832) and Alroy (2008).

Fossil distribution

- Doue-la-Fontaine France estimated age: ~13.65—7.25 Mya.
- Lower Bakhtiari Formation, northern Iraq, estimated age: ~11.6—9.0 Mya.
- Kurtchuk-Tchekmedje, Turkey estimated age: ~11.61—5.33 Mya.
- Lang. E Quarry-GM aka ETE Locality 1042, Langebaanweg, "E Quarry, GM", South Africa.

H. quinni

Cormohipparion quinni is exclusive to the Great Plains of North America and was named by Woodburne in 1996. Its type locality is Devil's Gulch Horse Quarry, which is in a Barstovian terrestrial horizon in the Valentine Formation of Nebraska. It was recombined as *Hippotherium quinni* by Alroy (2007).

Body mass

A single specimen was examined by M. Mendoza, C. M. Janis, and P. Palmqvist for body mass. (Mendoza et al).

- Specimen: 79.9 kg (180 lb).
- Kennesaw site, Logan County, Colorado estimated age: ~12.9 Mya.
- Devil's Jump Off, Nenzel Quarry, and Sawyer Quarry, and Valentine Railway Quarries, Cherry County, Nebraska 13-12.9 Mya.
- Schoettger Quarry, Keya Paha County, Nebraska 12.9-12.7 Mya.

Source (edited): "http://en.wikipedia.org/wiki/Hippotherium"

Hispanotherium

Hispanotherium was a genus of rhinoceros of the tribe Elasmotheriini endemic to Europe and Asia during the Miocene living from 16—7.25 mya existing for approximately 8.75 million years.

Taxonomy

Hispanotherium was named by Crusafont and Villalta (1947). Its type is *Hispanotherium lintungensis*. It was assigned to Rhinocerotidae by Carroll (1988); to Elasmotheriini by Deng (2003); and to Iranotheriinae by Guérin and Pickford (2003).

Source (edited): "http://en.wikipedia.org/wiki/Hispanotherium"

Homalodotheriidae

Homalodotheriidae is an extinct family comprising four genera of notoungulate mammals known from the late Eocene through late Miocene of South America.

Source (edited): "http://en.wikipedia.org/wiki/Homalodotheriidae"

Homalodotherium

Homalodotherium is an extinct genus of the order Notoungulata, an extinct group of hoofed mammals native to South America.

Homalodotherium was about 2 metres (6.6 ft) in body length, and had long forelimbs with claws instead of hooves. It walked on the soles of its hind feet and the toes of its front feet, which would have made the animal higher at the shoulder than at the hips when it walked on all fours. It was probably at least partially bipedal, being able to pull down tree branches with its arms while rearing up on its hind legs. Various other prehistoric and living creatures have also developed this feeding style; examples are the chalicotheres, ground sloths, the giant panda and, possibly, the therizinosaur dinosaurs.

Source (edited): "http://en.wikipedia.org/wiki/Homalodotherium"

Homunculus patagonicus

Homunculus patagonicus is an extinct New World monkey species that lived in Argentina during the Miocene. It is likely the ancestral species for the Callicebinae subfamily, whose living members are the titi monkeys.

Source (edited): "http://en.wikipedia.org/wiki/Homunculus_patagonicus"

Honanotherium

Honanotherium schlosseri was a giraffid ancestral to the modern giraffe (genus *Giraffa*) from the late Miocene of Hunan Province, China. It would have resembled a modern giraffe, but, somewhat shorter.

Etymology

reconstruction

The first part of the generic name, *ho-nano* refers to the Hunan (Chinese: 湖南; pinyin: *Húnán*) province of China, where the first specimens were recovered. the second part, *therium*, comes from the Greek, *θηρίον* which means "beast." The species name, "schlosseri," honors M. Schlosser, a German paleontologist who specialized in prehistoric giraffes.
Source (edited): "http://en.wikipedia.org/wiki/Honanotherium"

Hoplitomeryx

The extinct five-horned prongdeer ***Hoplitomeryx matthei*** with its sabrelike ('moschid' type) upper canines lived on the former Gargano Island during the Miocene and the Early Pliocene, now a peninsula on the east coast of South Italy.

Its fossilized remains were retrieved in the late sixties and subsequent years from reworked reddish, massive or crudely stratified silty-sandy clays (terrae rossae), which partially fill the paleo-karstic fissures in the Mesozoic limestone substrate and that are on their turn overlain by Late-Pliocene-Early Pleistocene sediments of a subsequently marine, shallow water and terrigenous origin. In this way a buried paleokarst originated.

The fauna from the paleokarst fillings is known as *Mikrotia* fauna after the endemic murid of the region (initially named "Microtia", with a c, but later corrected, because the genus *Microtia* was already occupied). Later, after the regression and continentalization of the area, a second karstic cycle started in the late Early Pleistocene, the neokarst, which removed part of the paleokarst fill.

Appearance

Hoplitomeryx was a deer-like ruminant with a pair of pronged horns above each orbit and one central nasal horn. Hoplitomerycidae are not the only horned deer, before the appearance of antlered deer, members of the deer family had horns. Another left-over of this stage is Antilocapra of North America, only survivor of a once successful group related to Bovidae.

The diagnostic features of *Hoplitomeryx* are: one central nasal horn and a pair of pronged orbital horns, protruding canines, complete fusion of the navicocuboid with the metatarsal, distally closed metatarsal gully, a non-parallel-sided astragalus , and an elongated patella.

Hoplitomeryx species

The *Hoplitomeryx* skeletal material forms a heterogeneous group, containing four size groups from tiny to huge; within the size groups different morphotypes may be present. All size groups share the same typical *Hoplitomeryx* features. The different size groups are equally distributed over the excavated fissures, and are therefore not to be considered chronotypes. The hypothesis of an archipelago consisting of different islands each with its own morphotype cannot be confirmed so far. The tiny and small specimens show insular dwarfism, but this cannot be said for the medium and huge specimens.

The situation with several co-existing morphotypes on an island is paralleled by *Candiacervus* (Pleistocene, Crete, Greece). Opinions about its taxonomy differ, and at present two models prevail: one genus for eight morphotypes, or alternatively, two genera for five species. The second model is based upon limb proportions only, but these are invalid taxonomic features for island endemics, as they change under influence of environmental factors that differ from the mainland. Also in *Hoplitomeryx* the morphotypes differ in limb proportions, but here different ancestors are unlikely, because in that case they all ancestors must have shared the typical hoplitomerycid features. Interestingly, in Candiacervus as well as in *Hoplitomeryx*, the largest species is as tall as an elk, but gracile and slender.

The large variation is instead explained as an example of adaptive radiation, starting when the Miocene ancestor colonized the island. The range of empty niches promoted its radiation into several trophic types, yielding a differentiation in *Hoplitomeryx*. The shared lack of large mammalian predators and the limited amount of food in all niches promoted the development of derived features in all size groups (apomorphies).

Source (edited): "http://en.wikipedia.org/wiki/Hoplitomeryx"

Horned gopher

Horned gophers are rodents from the genus ***Ceratogaulus***, a member of the extinct fossorial rodent family Mylagaulidae. *Ceratogaulus* is the only known rodent genus with horns, and is the smallest known horned mammal. *Ceratogaulus* lived from the late Miocene to the early Pleistocene era.

Description

Reconstruction of *Ceratogaulus hatcheri*

The horned gopher had two horns as shown to the left; these were large (in comparison to body size), paired, and originated from the nose. Horned gophers are the smallest known mammals to ever have horns, and the only known species of horned rodents, and (aside from one fossil species of armadillo, *Peltephilus*), the only known fossorial horned mammals. They were native to what is now the Great Plains of North America, most concentrated in Nebraska.

The role of the horns is subject to much speculation; possibilities include digging (although this has largely been ruled out by the horns' position and orientation; see below for a more detailed analysis), mating displays or combat, and defense from predators. Because the horns were not sexually dimorphic, their role in defense seems most likely.

In most other respects, the animals resembled modern marmots. They were approximately 30 centimetres (0.98 ft) long, and had paddle-like forepaws with powerful claws adapted for digging. They also had small eyes, and probably had poor eyesight, similar to that of a mole. These features suggest that they were likely to be burrowing animals.

Digging

It has been believed that the horns of the horned gopher were used for digging, but the nasal horns of Ceratogaulus are inconsistent with use as a digging tool. In recent mammals that use the head for excavating, the tip of the snout is used like a spade to scrape at the substrate Therefore, the only modification of the nasal bones is a slight thickening of the anterior tips. Although it is theoretically possible that some mammal might develop horns as a digging tool, digging horns would differ from the Ceratogaulus horns in position and shape. Ceratogaulus horns are positioned on the posterior ends of the nasal bones and extend dorsally, perpendicular to the plane of the palate. As a result of their posterior position, using the horns to dig would bring the anterior tip of the nasals against the substrate after a very short sweep of the horns, making digging with the horns extremely inefficient. This motion would be even more inefficient than suggested because the anterior surface of a burrow is concave, making it essentially impossible to use the horns without the anterior end of the snout interfering. The expectation is that an animal using its horns anteriorly (rather than dorsally) would have the occipital plate positioned vertically or tilted posteriorly. In this configuration, the effective input lever is maximized when the head is lowered, as in the rhinoceros skull.

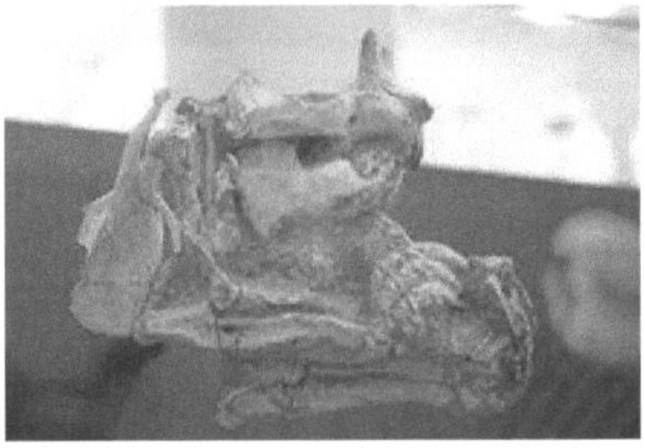

Ceratogaulus rhinocerus fossil

The shape of the horn itself is also very poor for a digging tool. The horns are very thick and broad with large, flat anterior and posterior surfaces. Dragging such a broad tool through the soil would create immense resistance, proportional to the large surface area presented to the substrate. Finally, the Ceratogaulus horn becomes more posteriorly positioned through time, so that the evolutionary trend is towards a horn which becomes more poorly suited to digging through time, rather than better suited. Thus, the argument that the horns functioned in digging is not supported by the morphology or the evolutionary progression.

Mating displays or combat

1902 Illustration of a pair of horned gophers

Many of the objections that apply to the horns as a digging implement also apply to the use of the horns in sexual combat. Their orientation and position and the morphology of the rest of the skull make it exceedingly difficult to bring them to bear on an opponent of similar size. The cervical vertebrae are shortened antero-posteriorly in all mylagaulids (a feature inherited by *Ceratogaulus* from ancestral, head-digging mylagaulids), decreasing the flexibility and range of motion of the neck and making it even more difficult for *Ceratogaulus* to wrestle with their horns. Many ungulates with horns ill-suited to sexual combat still use them for combat or for sexual display. However, a sexually selected use of the horns is unlikely in *Ceratogaulus*, as the optic foramen is very small, roughly 1/2 - 2/3 times the size of that of the mountain beaver, *Aplodonta rufa*, which itself has very poor vision. The small size of the optic foramen indicates extremely poor visual acuity, meaning that the females would be unlikely to be able to visually recognize a winner in any sexual displays or sexual combat by the males.

Defense

Horns are used in defense against predators by almost all horned mammals. Animals will use any weapons at their disposal to fight off predators, and the horns of Ceratogaulus are particularly well suited to defense. The horns are broad and robust and their dorsal orientation and relatively posterior position makes them well suited to protecting the vulnerable eyes and neck. By elevating the head dorsally, the horns would be snapped backward, protecting the areas most commonly attacked by predators. A similar use of postero-dorsal horns has been indicated to decrease predation in horned lizards. As the horns grow taller through evolutionary time, they also become more posteriorly positioned and the height of the occipital plate increases, increasing the leverage for lifting them. By positioning the horns more posteriorly, the output lever is shortened and, because the muscles used to rotate the skull dorsally attach at the top of the occipital plate, the input lever is lengthened. Thus, the dorsal strike with the horns would be more powerful as the ratio of output lever to input lever would be increased. This increased leverage would also be helpful in maneuvering the horns to deter predators from attacking at the head or neck, as elevating the head dorsally would snap the horns backward, protecting these vulnerable areas. Predation is by far the dominant cause of mortality in most small mammals, so the benefits provided by a mechanism to reduce predation could offset the substantial evolutionary cost of horns in a fossorial mammal.

Source (edited): "http://en.wikipedia.org/wiki/Horned_gopher"

Hyaenodon

Hyaenodon ("hyaena-toothed") is an extinct genus of Hyaenodonts, a group of carnivorous creodonts of the family *Hyaenodontidae* endemic to all continents except South America, Australia and Antarctica, living from 42—15.9 mya, existing for approximately 26.1 million years.

Morphology

Some species of this genus were amongst the largest terrestrial carnivorous mammals of their time; others were only of the size of a marten. *Hyaenodon* was one of the latest genera of the Hyaenodonts and is known from the Late Eocene to Early Miocene. Remains of many species are known from North America, Europe, Asia and Africa (In 1993 42 species were distinguished).

Typical of early carnivorous mammals, the Hyaenodon had a very massive skull but only a small brain. It had a long skull with a narrow snout - much larger in relation to the length of the skull than in canine carnivores, for instance. Its neck was shorter than its skull, while its body was long and robust and terminated in a long tail. Despite the name, these creatures are not related to hyenas.

The average weight of adult or subadult *H. horridus*, the largest North American species, is estimated to about 40 kilograms (88 lb) and may not have exceeded 60 kilograms (130 lb). *H. gigas*, the largest *Hyaenodon* species was much larger, being 500 kilograms (1,100 lb) and around 10 feet. *H. crucians* from the early Oligocene of North America is estimated to only 10 to 25 kilograms (22 to 55 lb). *H. microdon* and *H. mustelinus* from the late Eocene of North America were even smaller and weighed probably about 5 kilograms (11 lb).

Reconstruction by Heinrich Harder, c.1920

Range and species

In North America the last *Hyaenodon* disappeared along with species like *H. brevirostris* in the late Oligocene. In Europe they had already vanished earlier in the Oligocene. From the Miocene in Africa there are three species (*H. andrewsi*, *H. matthewi* and *H. pilgrimi*) known, but none of these reached the dimensions of Asian species like *H. gigas* and *H. weilini*

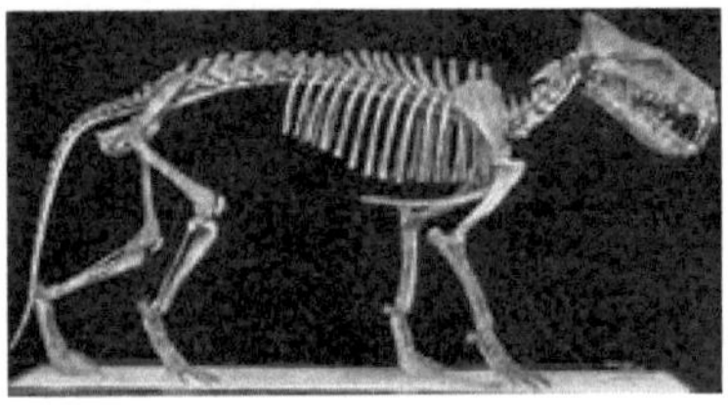

Hyaenodon horridus skeleton.

Popular culture

It appeared in Walking with Beasts attacking an *Indricotherium's* calf. It is then shown killing a Chalicotherium but the kill is stolen by a trio of Entelodons. Then, one more is shown chasing an entelodon.

David Innes adopts a hyenadon as a personal pet, naming him Raja, in Pellucidar, the second book of Edgar Rice Burroughs' Earth's Core series.

In *The Land That Time Forgot* (Caspak, #1), also by Edgar Rice Burroughs, Lys is found confronted by three hyaenodons, having apparently killed another, presumably by a lucky shot.

In the *National Geographic* series "Prehistoric Predators", episode Razor Jaws the creature was featured as a top predator yet competing against the Saber toothed cat, Dinictis, and Daeodon.But,would be eventually out competed by Amphicyon the bear dog.

In the "Paleoworld" episode Valley of The Uglies the hyaenadon was said to be able to break the neck of any creature it's size and rip the hide of an alligator.

Several hyenadons appeared in Episode 4.6 of the ITV science-fiction drama *Primeval.*

Source (edited): "http://en.wikipedia.org/wiki/Hyaenodon"

Hypertragulidae

Hypertragulidae is an extinct family of even-toed ungulates (order Artiodactyla), endemic to North America, Europe, and Asia during the Eocene through Miocene, living 46.2—13.6 Ma, existing for approximately 32.6 million years.

Hypertragulidae were primitive and ancient ruminants, resembling small deer or musk deer. However, both deer and musk deer are no longer considered to be closely related to the hypertragulids. Instead, the chevrotains are probably the closest living relatives to these ancient deer-like animals.

Taxonomy

Hypertragulidae was named by Cope (1879). It was considered paraphyletic by Matthew (1908). It was assigned to Ruminantia by Matthew (1908) and Gregory (1910); to Pecora by Cook (1934); and to Traguloidea by Carroll (1988).

Morphology

Hypertragulidae have tetradactyl front feet and ditactyl (two digits) rear feet which is specific to this family and no other ruminant. They ranged in body mass from as small as 2.16 kg (4.8 lb) with Parvitragulus to as large as 424.9 kg (940 lb) in Hypisodus.

Source (edited): "http://en.wikipedia.org/wiki/Hypertragulidae"

Hypertragulus

Hypertragulus is an extinct genus belonging to the family Hypertragulidae, within the order Artiodactyla, endemic to North America during the Eocene to Miocene, living 42 to 20.6 million years ago, existing for approximately 21.4 million years.

Hypertragulus were primitive and ancient ruminants, resembling small deer or musk deer, although they were more closely related to modern chevrotains. Its diet is stated to be that of a frugivore.

Morphology

One specimen of *Hypertragulus* examined for body mass by M. Mendoza, C. M. Janis and P. Palmqvist had a mass of 7.29 kg (16 lb).

Fossil distribution

A partial list of fossil sites:

- Chihuahua, Mexico
- Cedar Creek Formation (Lower & Middle), Logan County, Colorado
- Fort Logan Formation, Meagher County, Montana
- Upper Pomerado Conglomerate Formation, San Diego County, California

Source (edited): "http://en.wikipedia.org/wiki/Hypertragulus"

Hypohippus

Hypohippus ("Low Horse") is an extinct genus of three-toed Horse, which lived 17 - 11 million years ago. It was the size of a pony, and fossils of it have been found in Nebraska, Colorado, and Montana.

Restoration

Source (edited): "http://en.wikipedia.org/wiki/Hypohippus"

Hypsiops

Hypsiops is an extinct genus of oreodont of the superfamily *Merycoidodontoidea*, family *Merycoidodontidae* endemic to North America during the Late Oligocene-Early Miocene epochs (24.8—16.3 mya), existing for approximately 8.5 million years.

Taxonomy

Hypsiops was named by Schultz and Falkenbach (1950). It was assigned to *Merycoidodontidae* by Schultz and Falkenbach (1950) and Lander (1998).

Morphology

Two specimens were examined by M. Mendoza for body mass. Specimen one was estimated to have a weight of 168.6 kg (371.6 lbs). Specimen two was estimated to have a weight of 223.0 kg (491.6 lbs).

Fossil distribution

Fossils have been uncovered throughout the western U.S. in Oregon, Montana, Wyoming, as well as Nebraska.

Species

H. bannackensis (syn. *Pseudomesoreodon boulderensis, Pseudomesoreodon rolli*), *H. breviceps* (syn. *H. johndayensis, Pseudomesoreodon rooneyi, Ticholeptus brachymelis, Ticholeptus petersoni*.

Source (edited): "http://en.wikipedia.org/wiki/Hypsiops"

Hyracodontidae

Hyracodontidae is an extinct family of rhinoceroses endemic to North America, Europe, and Asia during the Eocene through early Miocene living from 55.8—20 mya, existing for approximately 35.8 million years.

They are typified as having long limbs and having no horns. These animals were initially modest in size and fast moving, having evolved from smaller members of Rhinocerotoidea during the Late Eocene and Early Oligocene. They later evolved into gigantic forms that included the largest terrestrial mammals ever to have lived (the Indricotheriinae or Paraceratheriinae).

Hyracodonotidae thrived in the rainforests of Kazakhstan, Pakistan, and southwest China, a former coastal region.

Source (edited): "http://en.wikipedia.org/wiki/Hyracodontidae"

Ictiocyon

Ictiocyon is an extinct genus of large, mostly carnivorous bone-crushing mammals known as bear dogs, of the family Amphicyonidae endemic to Asia during the Miocene living from 23.03—15.97 Ma and existed for approximately 7.06 million years.

Taxonomy

Ictiocyon was named by Crusafont Pairó et al. (1955). It was assigned to Amphicyonidae by Wang et al. (2005).

Morphology

A single specimen was examined by Legendre and Roth for body mass and was estimated to weigh 1.42 kg (3.1 lb).

Source (edited): "http://en.wikipedia.org/wiki/Ictiocyon"

Ictitherium

Ictitherium is an extinct genus belonging to the family Hyaenidae and the subfamily Ictitheriinae erected by Trouessart in 1897. *Ictitherium* species were endemic to Eurasia and Africa during the Middle Miocene through the Early Pliocene (12.7—5.3 mya) and existed approximately 7.4 million years.

Ictitherium were around 1.2 metres

(4 ft) long, and looked more like civets than modern hyenas, possessing a long body with short legs. Judging from the teeth, they were insectivores. *Ictitherium* was a very successful and abundant genus, with multiple fossils often often being found at a single site. Possibly, this early hyena genus lived in packs and had a social order, much like its modern descendants.

Species

- *I. viverrinum*
- *I. syvalense*
- *?I. arambourgi*
- *?I. gaudryi*
- *?I. hipparionum*
- *?I. orbingyi*
- *?I. sinence*
- *?I. preforfex*

Source (edited): "http://en.wikipedia.org/wiki/Ictitherium"

Ilingoceros

Ilingoceros is an extinct genus of pronghorn from the Miocene of North America.

At 1.8. metres (5.9 ft) in body length, the animal would have been slightly bigger than the modern pronghorn. It had straight, spiraled horns, which ended in forked tips.

Source (edited): "http://en.wikipedia.org/wiki/Ilingoceros"

Imagocnus

Imagocnus is an extinct genus of ground sloth from the early Miocene of Cuba.

This sloth resided in the Antilles and showed a range of body size. Its relationships to other Antillean sloths are not immediately clear, though the genera *Megalocnus* and *Parocnus*,other ground sloths, are its most likely relatives.

Source (edited): "http://en.wikipedia.org/wiki/Imagocnus"

Imagotaria

Imagotaria is an extinct monotypic genus of walrus with the sole species ***Imagotaria downsi***. Fossils of *Imagotaria* are known from the early late Miocene of California (ca. 10-12 million years ago).

Description

The 1.8 metres (6 ft) long pinniped more closely resembled in its overall shape a sea lion rather than a walrus. Unlike the extant walrus (*Odobenus rosmarus*), *Imagotaria* did not possess elongate, ever-growing tusks, but instead bore enlarged canines (with respect to other pinnipeds).

Imagotaria is a example of a primitive walrus that does not grossly appear similar to a modern walrus. However, the walrus family (the Odobenidae) is a more inclusive group, that includes walruses without tusks (e.g. *Imagotaria*), walruses with upper and lower tusks (the subfamily Dusignathinae), and walruses with upper tusks like the extant walrus (subfamily Odobeninae, tribe Odobenini). It is possible to classify these pinnipeds as walruses because they share many other skull features (besides tusks) as well as many skeletal features, all of which indicate common ancestry.

Palaeobiology

The teeth of *Imagotaria* indicate that its feeding ecology was markedly different from that of modern walrus, and more similar to that of less specialized pinnipeds like seals, fur seals, and sea lions. Conical, unworn teeth and the lack of a vaulted palate indicate that *Imagotaria* did not feed on molluscs like modern walrus. Modern walruses do not use their teeth to chew molluscs like sea otters do. Instead, they hold a clam in their lips, and the vaulted palate allows them to use their tongue as a powerful piston to suck the soft parts right out of the clam shell. The shell is then dropped to the seafloor, never entering the oral cavity.

Additionally, fossils of *Imagotaria* (and the earlier *Neotherium*, ca. 15 million years ago) demonstrate that early walruses had, by the middle and late Miocene, already developed extreme sexual dimorphism (males and females having different body sizes). It is unclear whether extreme sexual dimophism is ancestral to all pinnipeds, or if it has been independently acquired in multiple pinniped lineages.

Source (edited): "http://en.wikipedia.org/wiki/Imagotaria"

Indarctos

Indarctos is a genus of mammals of the family Ursidae (bear) endemic to North America, Europe, Africa, and Asia during the Miocene, living from ~11.1—5.3 Ma, existing for approximately 6.2 million years.

The oldest member is from Arizona (~11.1—7.7 Ma.) and youngest is (~9.0—5.3 Ma.) from Kazakstan. In

North America this animal was contemporary with *Plionarctos* (~10.3—3.3 Ma).

Taxonomy

Indarctos was named by Pilgrim (1913) Its type is *Indarctos salmontanus*. It was assigned to Agriotheriini by Chorn and Hoffman (1978); to Ursavini by Hunt (1998); and to Ursidae by Pilgrim (1913), Carroll (1988) and Salesa et al. (2006).

Morphology

Body mass

Two specimens were examined by Legendre and Roth for body mass.

- Specimen 1 estimated to weigh: 244.4 kg (540 lb)
- Specimen 2 estimated to weigh: 2,517.4 kg (5,500 lb)

Fossil distribution

Sites and specimen ages:

- Box T Site, Lipscomb County, Texas ~9.3—9.2 Ma.
- Rattlesnake site, Grant County, Oregon ~10.3—4.9 Ma.
- Withlacoochee River Site 4A, Marion County, Florida paleontological sites *(Indarctos sp.)* ~10.3—4.9 Ma.
- Lufeng, Yunnan, China *(I. atticus)* ~9—5.3 Ma.
- Yulafli (CY), Thrace, Turkey *(I. arctoides)* ~9.7—8.7 Ma.

Species

- *I. nevadensis*, and *I. oregonensis* are of North America with fossil recovery in Oregon and Nevada, USA
- *Indarctos salmontanus* (Pilgrim, 1913) is the species type originating in Calcutta, India and discovered by the Geological Survey of India.

Source (edited): "http://en.wikipedia.org/wiki/Indarctos"

Interatheriidae

Interatheriidae is an extinct family of notoungulate mammals from South America. Interatheriids are known from the Paleocene or Eocene through the Miocene (McKenna and Bell, 1997).

Source (edited): "http://en.wikipedia.org/wiki/Interatheriidae"

Interatherium

Interatherium is an extinct genus of Interatheriidae from the early Miocene.

Source (edited): "http://en.wikipedia.org/wiki/Interatherium"

Iranotherium

Iranotherium ("Beast of Iran") was a large elasmothere rhinoceros, as big as a modern white rhino, found in Central Asia. It was a precursor of the related *Sinotherium*, and may have been ultimately outcompeted by its descendant.

Skull of *Iranotherium morgani*.

Males were larger than the females and had more developed muscle scars on the zygomatic arch.

Source (edited): "http://en.wikipedia.org/wiki/Iranotherium"

Ischyrocyon

Ischyrocyon is an extinct genus of large, mostly carnivorous mammals (bone-crushers) known as bear dogs, of the family Amphicyonidae, subfamily Amphicyoninae, which inhabited North America during the Miocene living from ~16.3—10.3 Ma and existed for approximately 6 million years.

Taxonomy

Ischyrocyon was named by Matthew and Gidley (1904). Its type is Ischyrocyon hyaenodus. It was assigned to Amphicyonidae by Matthew and Gidley (1904) and Carroll (1988); and to Amphicyoninae by Hunt (1998).

Morphology

A single specimen was examined by Legendre and Roth for body mass.

- 161.9 kg (360 lb).

Fossil distribution

- Blair Junction, Esmeralda County, Nevada ~14.2—11.7 Ma.
- Hemicyon Quarry Site, San Bernardino County, California ~14.0 Ma.
- Adam Risley Ranch Site, Donley County, Texas ~12—11.7 Ma.
- WaKeeney (Unit C) Trego County, Kansas ~11.8—11.7 Ma.
- Olcott Hill, Sioux County, Nebraska ~11.9 Ma.
- Iron Canyon Site, Kern County, California ~11.8—11.9 Ma.
- Bluejay Quarry Site, Antelope County, Nebraska ~11.8 Ma.

Source (edited): "http://en.wikipedia.

org/wiki/Ischyrocyon"

Kanuites

Kanuites is an extinct genus of viverrid native to Africa.

Kanuites was about 90 centimetres (3 ft) long, and looked remarkably similar to modern genets. *Kanuites* was probably an omnivore and may have had retractable claws, like a feline. It may have lived at least part of its life in trees.

Source (edited): "http://en.wikipedia.org/wiki/Kanuites"

Kentriodon

Kentriodon is an extinct species of toothed whale. Fossils have been found in Europe, Asia, New Zealand, and the United States.

Kentriodon was the most diverse of all the kentriodontids, which include three named species and five undescribed species. These were small to medium-sized odontocetes with largely symmetrical skulls, and thought likely to include ancestors of some modern species. *Kentriodon* is also the oldest described kentriodontid genus, reported from the Late Oligocene to the Middle Miocene.

Kentriodontines ate small fish and other nectonic organisms; they are thought to have been active echolocators, and might have formed pods. The diversity, morphology and distribution of fossils appear parallel to some modern species.

Source (edited): "http://en.wikipedia.org/wiki/Kentriodon"

Kentriodontidae

Kentriodontidae is an extinct family of odontocet whales related to modern dolphins. The Family lived from the Oligocene to the Pliocene before going extinct.

Taxonomy

The kentriodontids are grouped in three subfamilies with two genera considered incertae sedis, or of uncertain placement in the family.

- Family Kentriodontidae
  - Subfamily Kentriodontinae
    - *Belonodelphis*
    - *Delphinodon*
    - *Incacetus*
    - *Kentriodon* (syn. Grypolithax)
    - *Macrokentriodon*
    - *Microphocaena*
    - *Rudicetus*
    - *Tagicetus*
  - Subfamily Lophocetinae
    - *Hadrodelphis*
    - *Liolithax*
    - *Lophocetus*
  - Subfamily Pithanodelphininae
    - *Atocetus*
    - *Leptodelphis*
    - *Pithanodelphis*
    - *Sophianacetus* (syn. Mediocris)
  - Subfamily incertae sedis
    - *Sarmatodelphis*
    - *Kampholophos*

Source (edited): "http://en.wikipedia.org/wiki/Kentriodontidae"

Kenyapithecus

Kenyapithecus wickeri was a fossil ape discovered by Louis Leakey in 1961 at a site called Fort Ternan in Kenya. The upper jaw and teeth were dated to 14 million years ago. One theory states that *Kenyapithecus* may be the common ancestor of all the great apes. More recent investigations suggest *Kenyapithecus* is more primitive than that and is only slightly more modern than when *Proconsul* is considered to be an ape.

Evidence suggests that *Kenyapithecus wickeri* was one of the species that started a radiation of apes out of Africa.

Morphology

Kenyapithecus wickeri fossils, Muséum national d'histoire naturelle, Paris

Impressed by *Kenyapithecus*'s modern-looking teeth, Leakey declared *Kenyapithecus* to be "a very early ancestor of man himself."

Kenyapithecus possessed craniodental adaptations for hard object feeding including thicker molar enamel, and a large mandible, large premolars and upper incisors that are similar to those seen in living pitheciine monkeys *Kenyapithecus* also possessed macaque-like limbs adapted for a knuckle-walking mode of semi-terrestrial locomotion This could show that as hominins evolved, they passed through a knuckle-walking phase.

Kenyapithecus wickeri has very distinct features, especially details in the canine teeth and is similar to modern apes.

Source (edited): "http://en.wikipedia.org/wiki/Kenyapithecus"

Kenyapotamus

Kenyapotamus is a possible ancestor of living hippopotamids that lived in Africa roughly 16 million to 8 million years ago during the Miocene epoch. Its name reflects that its fossils were first found in modern-day Kenya.

Although little is known about the *Kenyapotamus*, its dental pattern bore similarities to that of the genus *Xenohyus*, a European tayassuid from the Early Miocene. This led some scientists to conclude that hippopotami were most closely related to modern peccaries and pigs.

Recent molecular research has suggested that hippopotamids are more closely related to cetaceans than to other artiodactyls. A morphological analysis of fossil artiodactyls and whales, which also included *Kenyapotamus*, strongly supported a relationship between hippopotamids and the anatomically similar family Anthracotheriidae. Two archaic whales (*Pakicetus* and *Artiocetus*) formed the sister group of the hippopotamid-anthracotheriid clade, but this relationship was weakly supported.

Source (edited): "http://en.wikipedia.org/wiki/Kenyapotamus"

Killikaike blakei

Killikaike blakei is an extinct species of New World monkey that lived in Argentina during the Early Miocene.

Killikaike blakei was collected in southern-most Argentina in January, 2005 and the type specimen consists of a remarkably well preserved face. Unfortunately the neurocranium is not present. The specimen was named for the locality where it was found Killik Aike Norte, on the estate of the Blake family.

Source (edited): "http://en.wikipedia.org/wiki/Killikaike_blakei"

Kipsigicerus

Kipsigicerus is an extinct genus of East African antelope from the Middle Miocene. Its closest living relative is the four-horned-antelope.

Source (edited): "http://en.wikipedia.org/wiki/Kipsigicerus"

Kolopsis

Kolopsis torus is an extinct genus and species of Diprotodontid marsupial from Alcoota Station in the Northern

Territory of Australia. Source (edited): "http://en.wikipedia. org/wiki/Kolopsis"

Kolponomos

Kolponomos is an extinct genus of marine bears which existed from the Hemingfordian age to the Aquitanian age of the Miocene epoch, about 20 million years ago. It was described in 1960 by Ruben A. Stirton, a paleontologist at the American Museum of Natural History, from a partial skull and jaw found on the Olympic Peninsula. Stirton thought the fossil was a large marine procyonid, but a specimen found in two pieces by fossil collector Douglas Emlong near Newport, Oregon, in 1969 and 1977, proved the genus was related to ancestors of bears. Since that time, one other *Kolponomos* fossil has been discovered, in the Aleutian Islands, bringing the total number of specimens to three.

Kolponomos had downturned snouts and broad, heavy molars that would have been suited to a diet of hard-shelled marine invertebrates, and their narrow snouts and anteriorly directed eyes indicate that they would have been able to view objects directly in front of their heads. Large neck muscle attachments and robust foot bones combine with these features to suggest that *Kolponomos* filled a unique niche among marine carnivores, approached today only by the unrelated sea otter. Due to the lack of a complete skeleton, however, it is difficult to make inferences about this group's other adaptations.

Source (edited): "http://en.wikipedia. org/wiki/Kolponomos"

Kolponomos clallamensis

Kolponomos clallamensis is an extinct species of marine bear from the Miocene epoch.

Source (edited): "http://en.wikipedia. org/wiki/Kolponomos_clallamensis"

Kolponomos newportensis

Kolponomos newportensis is an extinct species of marine bear which existed from the Hemingfordian age to the Aquitanian age of the Miocene epoch, about 20 million years ago. It was described in 1994 by R. Tedford, L. Barnes and Clayton E. Ray. It is represented by single specimen: a nearly complete skull, jaw and post-cranial bones found in a concretion of sediment. The concretion was discovered in two pieces by fossil collector Douglas Emlong near Newport, Oregon, the first in 1969 and the second, eight years later, in 1977. Because the concretion had been hardened so much by tectonic stress, the paleontological laboratory at the Smithsonian Institution considered them "the most difficult materials ever encountered by our laboratory.," and a combination of techniques proved essential to its extraction and preparation, which lasted two decades. Discovery of *K. newportensis* disproved the earlier theory that the genus was related to ancestors of raccoons, and showed instead that it is instead an early but unusual bear relative.

Source (edited): "http://en.wikipedia. org/wiki/Kolponomos_newportensis"

Kronokotherium

Kronokotherium is an extinct herbivorous marine mammal of the genus Desmostylia of the family *Desmostylidae. Kronokotherium* lived along the North American Pacific Coast from the Chattian stage of the Late Oligocene through the Aquitanian stage of the Early Miocene (28.4 mya—20.4 Mya) and existing for approximately 7.8 million years.

Taxonomy

Kronokotherium was named by Pronina (1957). Its type is Kronokotherium brevimaxillare.

Fossil distribution

Clallam County, Washington.

Species

K. brevimaxillare

Sister genera

Behemotops, Desmostylus, Cornwallius, Paleoparadoxia, Vanderhoofius.

Source (edited): "http://en.wikipedia. org/wiki/Kronokotherium"

Kubanochoerus

Kubanochoerus was a genus of large, long-legged pigs from the Miocene of Eurasia.

Description

The largest species, the aptly named *K.*

gigas, grew to be up 1.2 metres (3.9 ft) at the shoulder, and probably weighed up to 500 kilograms (1,100 lb) in life. The heads of these pigs were unmistakable, with small eyebrow horns, and a large horn emanating from the forehead of the males. It is speculated that the males used their forehead horns for jousting with each other.

The species *K. massai* was originally thought to be an African species belonging to this genus, as the first specimen shared the genus' distinctive eyebrow horns. Its lack of a forehead horn lead experts to interpret the skull as belonging to a female. However, recently, *K. massai* has been split off into its own genus, *Libyochoerus* (commemorating the fact that it was originally found in Libya.) Some species of the genus *Megalochoerus* have been also reassigned to *Libyochoerus*, though, many experts do not agree with this.

Source (edited): "http://en.wikipedia.org/wiki/Kubanochoerus"

Kyptoceratini

Kyptoceratini is an extinct tribe of the subfamily Synthetoceratinae, deer-like mammals within the family Protoceratidae belonging to the order Artiodactyla, endemic to North America during the Miocene through Pliocene, living 23.03—3.6 Ma, existing for approximately 19.43 million years.

Taxonomy

Kyptoceratini is a sister taxa to Synthetoceratini. *Kyptoceratini* was named by Webb (1981). Its type is Kyptoceras. It was assigned to Synthetoceratinae by Webb (1981), Prothero (1998), Webb et al. (2003) and Prothero and Ludtke (2007).

Members

Kyptoceras (type genus), Syndyoceras

Source (edited): "http://en.wikipedia.org/wiki/Kyptoceratini"

Lambdoceras

Lambdoceras is an extinct genus of Protoceratidae belonging to the order Artiodactyla (subfamily Synthetoceratinae) endemic to North America during the Miocene, living epoch 20.6—13.6 Ma, existing for approximately 7 million years.

Taxonomy

Lambdoceras was named by Stirton (1967). Its type is *Lambdoceras hessei*. It was synonymized subjectively with Prosynthetoceras by Patton (1969), Patton and Taylor (1971) and Patton and Taylor (1973). It was assigned to Protoceratidae by Stirton (1967), Tedford et al. (1987) and Albright (1999); and to Synthetoceratini by Webb (1981), Prothero (1998), Webb et al. (2003) and Prothero and Ludtke (2007).

Source (edited): "http://en.wikipedia.org/wiki/Lambdoceras"

Leontiniidae

Leontiniidae is an extinct family comprising six genera of notoungulate mammals known from the middle Eocene through middle Miocene of South America (McKenna and Bell, 1997).

Source (edited): "http://en.wikipedia.org/wiki/Leontiniidae"

Leptocyon

Leptocyon is a small extinct genus of canidae endemic to North America during the Oligocene through Miocene living from 24.8—10.3 mya, existing for approximately 15.5 million years.

Leptocyon was a small bodied, fox-like animal with a slender jaw.

Taxonomy

Leptocyon was named by Matthew (1918). It was synonymized subjectively with Vulpes by Gregory (1942); it was revalidated by Webb (1969), Carroll (1988) and Munthe (1998). It was assigned to Canidae by Matthew (1918), Webb (1969), Carroll (1988) and Munthe (1998).

Morphology

Body mass

Two specimens were examined by Legendre and Roth. Estimating the body mass of these two with both weighing an estimated 3.26 kg (7.2 lb)

Source (edited): "http://en.wikipedia.org/wiki/Leptocyon"

Leptomeryx

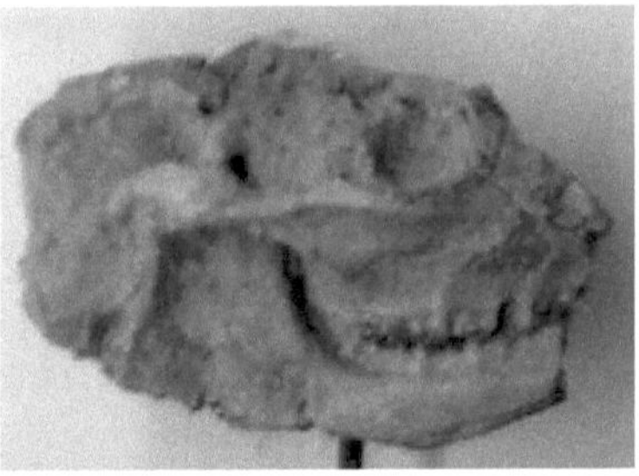

Leptomeryx evansi skull at the Museum für Naturkunde, Berlin

Leptomeryx is an extinct genus of ruminant of the family Leptomerycidae, endemic to North America during the Eocene through Oligocene 38—24.8 Mya, existing for approximately 13.2 million years.

Taxonomy

Leptomeryx was named by Leidy (1853). Its type is *Leptomeryx evansi*. It was assigned to Leptomerycinae by Matthew (1908); to Hypertragulidae by Cook (1934); and to Leptomerycidae by Leidy (1853) and then by Carroll (1988).

Morphology

Leptomeryx was a small deer-like ruminant with somewhat slender body.

Body mass

Four specimens were examined by Legendre and Roth for body mass.

- Specimen 1 weighed an estimated: 7.62 kg (17 lb)
- Specimen 2: 14.7 kg (32 lb)
- Specimen 3: 6.24 kg (14 lb)
- Specimen 4: 5.72 kg (13 lb)

Fossil distribution

Sites and species recovered:

- Titus Canyon, Inyo County, California *(L.blacki)* ~30.6—33.9 Ma.
- UNSM Sx-8 (Orella C), Sioux County, Nebraska *(L. elissae)* ~33.9—24.8 Ma.
- Anxiety Butte, Saskatchewan, Canada *(L. sp., L. evansi)* ~38—24.8 Ma.
- Calf Creek, Saskatchewan, Canada, *(L. mammifer)* ~38—33.3 Ma.
- Toadstool Park, Sioux County, Nebraska *(L. speciosus)* ~37.2—33.3 Ma.
- Medicine Pole Hills, Bowman County, North Dakota *(L. yoderi)* ~38—33.9 Ma.

Source (edited): "http://en.wikipedia.org/wiki/Leptomeryx"

Libycosaurus

Libycosaurus ("Lizard of Libya") was one of the last anthracothere genera. It lived from the Middle to the Late Miocene, and ranged throughout Central and Northern Africa, and in Uganda, in what was then a lush, marshy environment.

Source (edited): "http://en.wikipedia.org/wiki/Libycosaurus"

Libypithecus

Libypithecus markgrafi is an extinct genus of primate that lived in the Late Miocene.

Source (edited): "http://en.wikipedia.org/wiki/Libypithecus"

Lokotunjailurus

Lokotunjailurus

Life

Lokotunjailurus is the name of a genus of sabre-toothed cats which existed in Kenya and Chad during the Miocene epoch.

Source (edited): "http://en.wikipedia.org/wiki/Lokotunjailurus"

Longirostromeryx

Longirostromeryx is an extinct genus of saber-toothed deer.

Source (edited): "http://en.wikipedia.org/wiki/Longirostromeryx"

Macaca libyca

Macaca libyca was a prehistoric Macaque from the Late Miocene of Egypt

Source (edited): "http://en.wikipedia.org/wiki/Macaca_libyca"

Machairodontini

Machairodontini is an extinct tribe of large saber-toothed cats of the subfamily Machairodontinae, that lived in Europe, Asia, Africa and North America during the Miocene through Pleistocene living from 11.6 mya–11 000 years ago.

Taxonomy

Machairodontini was named by Gill (1872) as well as de Beaumont (1964). It was assigned to Machairodontinae by Berta and Galiano (1983).

Source (edited): "http://en.wikipedia.org/wiki/Machairodontini"

Machairodus

Machairodus was a genus of large machairodontine saber-toothed cats that lived in Europe, Asia, Africa and North America during the Miocene through Pleistocene living from 11.6mya—126,000 years ago, existing for approximately 11.474 million years.

Species

The fossil species assigned to the genus Machairodus were divided by Alan Turner in his book *The big cats and their fossil relatives* (1997) into two grades of evolutionary development - M. aphanistus and the North American "Nimravides" catacopis representing the more primitive grade, and M. coloradensis and M. giganteus representing the more derived grade. The characteristics of the more advanced grade include a relative elongation of the forearm and a shortening of the lumbar region of the spine to resemble that in living pantherine cats (Turner and Anton, 1997). These trends would be taken further in Homotherium, which is thought to have evolved from Machairodus.

Anatomy

"Nimravides" catacopis stood about 1 m at the shoulder. Machairodus coloradensis was apparently significantly larger, about 1.2 m at the shoulder, according to skeletal and life reconstructions in *The big cats and their fossil relatives* (Turner and Anton, 1997). If accurate, this would make M. coloradensis one of the largest felids.

The skull of Machairodus was noticeably narrow compared with the skulls of modern pantherine big cats, and the orbits were relatively small. The canines were long, thin and flattened from side to side but broad from front to back like the blade of a knife, as in Homotherium. The front and back edges of the canines were serrated when they first grew, but these serrations were worn down in the first few years of the animal's life.

There was marked sexual dimorphism in Machairodus giganteus, with males much larger than females (Turner and Anton, 1997).

A single fossil specimen was examined by Legendre and Roth and determined to have an approximate body mass of 201.8 kg.

Source (edited): "http://en.wikipedia.org/wiki/Machairodus"

Macrodelphinus

Macrodelphinus kelloggi was an orca-sized relative of the cetacean *Eurhinodelphis*. It lived off the coast of what is now California during the early Miocene. Because of its size, and inch-long teeth, it is believed to have been an apex predator. *Macrodelphinus* is known from a fragmentary skull from southern California.

Source (edited): "http://en.wikipedia.org/wiki/Macrodelphinus"

Mammacyon

Mammacyon is an extinct genus of large, mostly carnivorous bone-crushing mammals known as bear dogs, of the family Amphicyonidae endemic to North America during the Oligocene to Miocene living from 30.8—20.6 Ma and existed for approximately 10.2 million years.

Taxonomy

Mammacyon was named by Loomis (1936). Its type is Mammacyon obtusidens. It was assigned to Amphicyonidae by Loomis (1936) and Carroll (1988); and to Temnocyoninae by Hunt (1998).

Morphology

A single specimen was examined by Legendre and Roth for body mass and was estimated to weigh 39.3 kg (87 lb).

Source (edited): "http://en.wikipedia.org/wiki/Mammacyon"

Mediochoerus

Mediochoerus is an extinct genus of oreodont of the family *Merycoidodontidae*, subfamily *Merycoidodontinae*, endemic to North America during the Early Miocene-Middle Miocene subepochs (20.6—13.6 mya), existing for approximately 7 million years.

Taxonomy

Mediochoerus was named by Schultz and Falkenbach (1941) and assigned to *Merycoidodontidae* by Schultz and Falkenbach (1941) and Lander (1998).

Fossil distribution

Nebraska and California.

Species

M. blicki (type species), *M. johnsoni, M. mohavensis*

Source (edited): "http://en.wikipedia.org/wiki/Mediochoerus"

Megacamelus

Megacamelus is an extinct genus of terrestrial herbivore in the family Camelidae, endemic to North America from the Miocene through Pliocene 10.3—4.9 mya, existing for approximately 5.4 million years.

This was the largest species of Camelid to roam the Earth.

Taxonomy

Megacamelus was named by Frick (1929). It was assigned to Camelidae by Frick (1929) and Honey et al. (1998).

Morphology

Four specimens were examined for estimated body mass by M. Mendoza, C. M. Janis, and P. Palmqvist. These specimens were estimated to weigh:

- 3,695.7 kg (8,100 lb)
- 3,059.8 kg (6,700 lb)
- 2,246.8 kg (5,000 lb)
- 1,747.9 kg (3,900 lb)

Fossil distribution

Fossil distribution ranged from Nebraska to Idaho to Southern California.

Source (edited): "http://en.wikipedia.org/wiki/Megacamelus"

Megalibgwilia

Megalibgwilia is a genus of echidna known only from Australian fossils that incorporates the oldest known echidna species. It lived during the Pleistocene, becoming extinct about 50,000 years ago.

Megalibgwilia was first described from a broken left humerus by Richard Owen when he identified *M. ramsayi* in 1884. Complete skulls and postcranial fossils have since been described. A second species, *M. robusta*, was described in 1896 by Australian paleontologist William Sutherland Dun.

Although they are sometimes commonly referred to as giant echidnas, *Megalibgwilia* species are thought to have been similar in size to the contemporary Western Long-beaked Echidna, but with slightly longer forearms. They were smaller than a large species known from fossils in Australia, *Zaglossus hacketti*. *M. ramsayi* fossils have been found in deposits across mainland Australia and on Tasmania. *M. robusta* has only been found in New South Wales. *Megalibgwilia* was probably an insect-eater, like the Short-beaked Echidna, rather than a worm-eater like members of *Zaglossus*.

M. robusta is the oldest known echidna and the only known Miocene species.

Source (edited): "http://en.wikipedia.org/wiki/Megalibgwilia"

Megistotherium

Megistotherium (from Greek, *megistos* "greatest" + *therion* "beast" and *osteon* 'bone' + *thlaston* 'crushed, bruised' with *-es* being an agent noun: 'bone-crusher') is an extinct genus of creodonts, the only known species of which is ***Megistotherium osteothlastes***.. It is most likely a junior synonym of *Hyainailouros sulzeri*

Description

Megistotherium osteothlastes was an enormous hyaenodontid creodont that lived during the early Miocene Epoch some 23 million years ago. It has been found in the Ngorora Formation and Muruyur Formation Kenya, Egypt, Namibia, Uganda and Libya. Named by Robert Savage in 1973, it is one of the largest terrestrial carnivorous mammals known to have existed. It stood about 1.5 meters at the shoulders, 3.5 meters in the head and body length, 1 meter long tail and its skull was 65 cm long and weighing 500 kg

The carnassial teeth of *Megistotherium* (like those of other Creodonta) were the upper first molars, and overlapped with their lower molar counterparts like scissors to form a formidable and powerful shearing action. The land that is now the Sahara desert was much more fertile in the Miocene. A considerable amount of it was grassland and rainfall was plentiful. Lakes and ponds provided water for large fauna, which provided *Megistotherium* and other predators with an ample supply of prey. Large hyaenodontids like this one could have originally evolved as specialized predators or scavengers of large African herbivores. Mastodon bones have been found with its fossils, indicating that

Megistotherium may have hunted them for food.

Taxonomy

The order Creodonta comprised a diverse group of predators that were most successful during the Eocene before being ecologically displaced by the order Carnivora during the late Oligocene. Megistotherium emerged in the Miocene towards the end of the creodents flourishing; it was a part of a radiation of African hyaenodontids that occurred at that time. *Hyainailouros sulzeri* is very closely related to *Megistotherium*, extremely similar in size, structure and ratios - with a long tail, short limbs and robust body. Morales & Pickford (2005), Morlo, Miller & El-Barkooky (2007) and Morales, Pickford & Salesa (2008) suggested that *Megistotherium* is actually a junior synonym of *Hyainailouros sulzeri*, which is known by an almost complete skeleton, among other remains, and has been found in Europe, Asia and Namibia, and therefore comes from the same localities.

Source (edited): "http://en.wikipedia.org/wiki/Megistotherium"

Menoceras

Menoceras ("Crescent Horns") is a genus of extinct, small rhinoceros endemic to most of southern North America and ranged as far south as Panama during the early Miocene epoch. It lived from around 30.7—19.7 Ma, existing for approximately 11 million years.

Behaviour

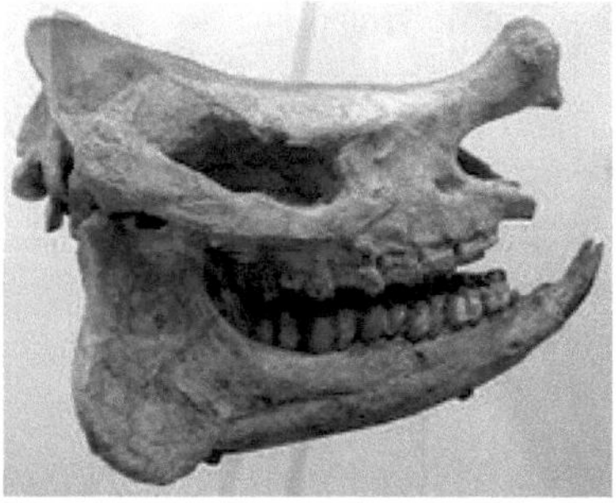

M. arikarense

M. cooki'

Menoceras roamed a grassland and plains environment of the North America (Prothero, 2005). Because of the massive accumulations of fossil bones of this animal, particularly at Agate Springs Nebraska, *Menoceras* may have lived large herds. However, other sites i.e. Martin-Anthony site Martin County, Florida, and Cady Mountains Horse Quarry, San Bernardino County, California.

Male *Menoceras* sported two horns side by side at the tip of the nose, whereas the females were hornless. All other rhino genera, save the related genus *Diceratherium*, have their horns arranged one behind the other. *Menoceros* grew to a length of 5 feet (1.5 meters) long, around the size of a large pig.

Taxonomy

Menoceras was named by Troxell and assigned to Rhinocerotidae by Troxell in 1921. It was synonymized subjectively with *Diceratherium* by Matthew in 1931 and Wood in 1964. Again assigned to Rhinocerotidae by Prothero, Guerrin, Manning in 1989. Tanner (1969), Wilson and Schiebout (1981), Prothero and Manning (1987), Carroll (1988) and Prothero et al. (1989); and to Menoceratinae by Prothero (1998).

Fossil distribution

Fossil distribution is as far north as New Jersey, south to Florida (3 collections) and Texas (6 collections), as far west as Nebraska (7 collections) and California (2 collections).

The Panamanian find was determined to be 19.7 Ma (AEO). It was found in the Gaillard Cut in Panama in "a 45 m thick section (narrow stratigraphic interval)" It was reposited in the Smithsonian Institute, Washington, D.C. Other sites:

- Lay Ranch Beds, Goshen County, Wyoming, a deposit of several carnivores and herbivores.
- Agate Springs Quarries, Sioux County, Nebraska, a deposit of Miocene herbivores and carnivores such as *Moropus elatus*, *Cynelos*, *Cephalogale*, and a number of Artiodactyla.
- Martin Canyon Quarry A, Logan County, Colorado, a very substantial number of carnivores and herbivores.

Morphology

Three specimens were examined for estimated body mass by M. Mendoza, C. M. Janis, and P. Palmqvist. These specimens were estimated to weigh:

- 91.7 kg (200 lb)
- 129.3 kg (290 lb)
- 482 kg (1,100 lb)

One specimen was examined by Mikael Fortelius, University of Helsinki and J. Kappelman and estimated a body mass of:

- 202.6 kg (450 lb)

Source (edited): "http://en.wikipedia.org/wiki/Menoceras"

Merychippus

Merychippus is an extinct proto-horse of the family Equidae that was endemic to North America during the Miocene from 20.43—10.3 Ma living for approximately 10.13 million years.

It had three toes on each foot and is the first horse known to have grazed. Its name means "ruminant horse", but current evidence does not support *Merychippus* ruminating.

Taxonomy

Merychippus was named by Leidy (1856). Numerous authors around the turn of the century put the type species in Protohippus, but this ignored priority. Its type is *Merychippus insignis*. It was assigned to Equidae by Leidy (1856) and Carroll (1988); and to Equinae by MacFadden (1998) and Bravo-Cuevas and Ferrusquía-Villafranca (2006).

Morphology

Body mass

Four specimens were examined by M. Mendoza, C. M. Janis, and P. Palmqvist for body mass. The results were:

- Specimen 1: 164.7 kg (360 lb)
- Specimen 2: 106.4 kg (230 lb)
- Specimen 3: 58.9 kg (130 lb)
- Specimen 4: 17.2 kg (38 lb)

Description

Merychippus lived in herds. It was about 89cm (35in) tall; at the time it was the tallest equine to have existed. The muzzle was longer, the jaw deeper, and the eyes wider apart than any other horse-like animal to date. The brain was also much larger, making it smarter and more agile. *Merychippus* was the first equine to have the distinctive head of today's horses.

The foot was fully supported by ligaments, and the middle toe developed into a hoof, which did not have a pad on the bottom. In some *Merychippus* species, the side toes were larger, whereas in others, they had become smaller and only touched the ground when running. Its teeth were like those of *Parahippus*; the extra crest that was variable in *Miohippus* was permanent in *Merychippus*, and the other teeth were beginning to form a series of tall crests with higher crowns.

Descendants

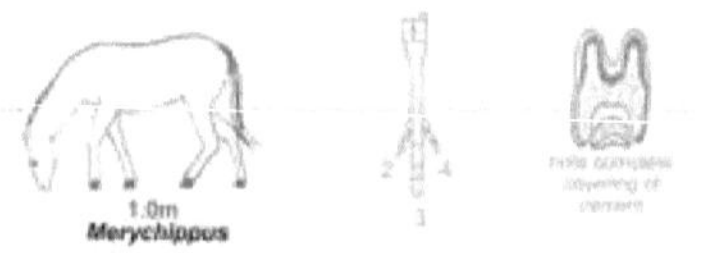

Illustration

By the end of the Miocene era, *Merychippus* was one of the first speedy grazers. It gave rise to at least nineteen different species of grazers, which can be categorized into three major groups. This burst of diversification is often known as the "Merychippine radiation."

The first was a series of three-toed grazers known as hipparions. These were very successful and split into four genera and at least sixteen species, including small and large grazers and browsers with large and elaborate facial fossae. The second was a group of smaller horses, known as protohippines, which included *Protohippus* and *Calippus*. The last was a line of "true equines" in which the side toes were smaller than those of other proto-horses. In later genera, these would be lost altogether as a result of the development of side ligaments that helped stabilize the middle toe during running.

Source (edited): "http://en.wikipedia.org/wiki/Merychippus"

Merycopotamus

Merycopotamus is an extinct genus of Asian anthracothere that appeared during the Middle Miocene, and died out in the Late Pliocene. At the height of the genus' influence, species ranged throughout southern Asia. With the extinction of the last species, *M. dissimilis*, the lineage of anthracotheres came to an end. *Merycopotamus* was closely related to the anthracothere genus *Libycosaurus*, which, unlike the former, never left Africa. In fact, some African fossils originally placed in *Merycopotamus*, but are now referred to *Libycosaurus*.

Source (edited): "http://en.wikipedia.org/wiki/Merycopotamus"

Mesocyon

Mesocyon is an extinct genus of terrestrial carnivore within the family Canidae, subfamily *Hesperocyoninae*, which inhabited North America from the Early Oligocene 33.9—20.6 Ma, existing for approximately 18 million years.

Though a carnivore, dentition suggests this animal was a hypercarnivore or mesocarnivore.

Taxonomy

Mesocyon was the most basal member of a clade that also includes the genera *Cynodesmus*, *Sunkahetanka*, *Philotrox*, and *Enhydrocyon*. Mesocyon was named by Scott (1890). It was synonymized subjectively with Enhydrocyon by Swisher (1982). It was assigned to Canidae by Scott (1890), Carroll (1988) and Stevens (1991).

Morphology

Two specimens were examined by Legendre and Roth with the first specimen's mass being estimated at 7.4 kg (16.3 lbs). The second specimen's mass was estimated at 7.19 kg (15.8 lbs).

Fossil distribution

Fossils were uncovered on 2 sites in north central Oregon. Also, 10 other sites from southern California Colorado, southern Wyoming, and western Nebraska.

Source (edited): "http://en.wikipedia.org/wiki/Mesocyon"

Mesopithecus

Mesopithecus ("middle monkey") is an extinct genus of Old World monkey that lived in Europe and western Asia 7 to 5 million years ago. It was once thought that it might be an ancestor of the grey langur, but a more recent study suggests that they are more closely related to the snub-nosed monkeys and doucs.

Mesopithecus resembled a modern macaque, with a body length of about 40 centimetres (16 in). It was adapted to both walking and climbing, possessing a slender body with long, muscular limbs and flexible fingers. Its teeth suggest that it primarily ate soft leaves and fruit.

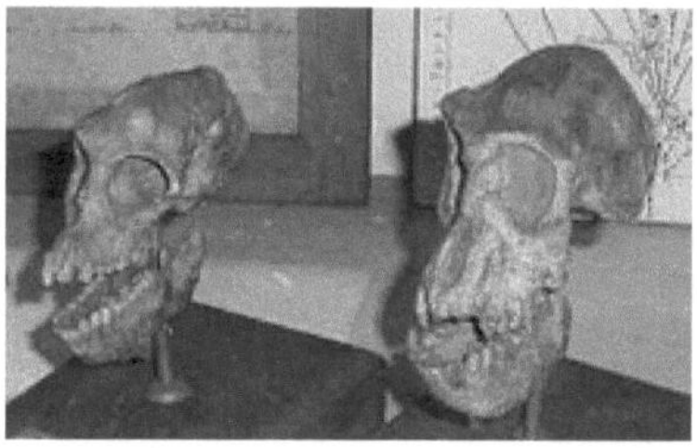

Mesopithecus pentelici skulls

Source (edited): "http://en.wikipedia.org/wiki/Mesopithecus"

Mesoreodon

Mesoreodon is an extinct genus of terrestrial herbivore of the family Merycoidodontidae, subfamily Merycoidodontinae (the oreodonts), endemic to North America during the Whitneyan stage of the Oligocene-Miocene epochs (33—20.6 mya) existing for approximately 12.4 million years.

Taxonomy

Mesoreodon was named by Scott (1893) and its type is *Mesoreodon chelonyx*. It was synonymized subjectively with *Eporeodon* by Lander (1998). It was assigned to Merycoidodontidae by Stevens and Stevens (1996).

Morphology

It was a large animal and ate the numerous low-growing plants and early grasses that sprung up on the plains of North America. It had a rather robust jaw, and like all oreodonts, sharp canine teeth.

Unlike many other oreodonts, who were restricted to certain habitats and places, *Mesoreodon* seemed to have been a cosmopolite. Fossils of *Mesoreodon* have been found in the Miocene deserts of California, the prairies of Nebraska, Wyoming, and South Dakota, southeastern Idaho, John Day Fossil Beds in Oregon, and Florida. In fact, *Mesoreodon* is the only Florida oreodont known from a complete skeleton, odd, since oreodonts had a continent-wide dominance until their extinction. They lived in all environments, and must have been very adaptable.

Mesoreodon had ossified vocal cords; the only other animal to have these in modern times is the howler monkey. *Mesoreodon* may have been a "screaming oreodont" using loud noises to intimidate its enemies and rivals.

Four specimens were examined by M. Mendoza for body mass and estimated to have a weight of:

- Specimen 1: 228.6 kg (504 lbs).
- Specimen 2: 162.2 kg (357.5 lbs).
- Specimen 3: 144.8 kg (319.2 lbs).
- Specimen 4: 202.4 kg (446.2 lbs).

Fossil distribution

Fossils were uncovered from Florida to California, Oregon, Montana, and Nebraska.

Source (edited): "http://en.wikipedia.org/wiki/Mesoreodon"

Mesotheriidae

Mesotheriidae ("Middle Beasts") is an extinct family of notoungulate mammals known from the Eocene through the Pleistocene of South America. Mesotheriids were small to medium-sized herbivorous mammals adapted for digging.

Characteristics

Mesotheriids were small to medium sized notoungulates; larger forms were approximately the size of a sheep (Shockey et al., 2007). Additionally, the family is characterized by specializations of the teeth and skeleton. In the dentition, all mesotheriids have ever-growing incisors with enamel restricted to the anterior surface, a condition termed gliriform, as it also occurs in Glires (rodents and lagomorphs). The cheek teeth (premolars and molars) of mesotheriids are high-crowned (hypsodont) and in advanced members of the family, the cheek teeth are also ever-growing (Shockey et al., 2007). Mesotheriid skeletons are heavily built and show features associated with digging in living mammals. In particular, fossorial characteristics of mesotheriids include deeply fissured claws, presence of a sesamoid bone in the elbow and reinforcement of the pelvic girdle by addition of vertebrae to the sacrum and fusion of the sacrum and innominate (Shockey et al., 2007).

Behavior

A biomechanical study of the skeleton of three mesotheriid genera (*Trachytherus*, *Plesiotypotherium*, and *Mesotherium*) spanning the temporal range of the family indicates that most or all mesotheriids were adapted for digging (Shockey et al., 2007). Shockey et al. concluded that mesotheriids likely dug for roots and tubers and were most similar in their diet and behavior to living wombats, although no living group is perfectly analogous. Extensive burrowing was considered possible but unlikely given the relatively large size of most mesotheriids.

Geographic and Temporal Distribution

As with almost all other notoungulates, mesotheriids are known only from the Cenozoic of South America (McKenna and Bell, 1997). Unlike some other families, mesotheriid fossils are not found across the continent. Instead, mesotheriids are most abundant and diverse in faunas from middle latitudes in Bolivia and Chile, particularly the Altiplano (Flynn et al., 2005). Mesotheriids fossils are rare in high latitude Patagonian faunas and absent entirely from tropical faunas in northern South America.

The earliest potential record of a mesotheriid is *?Trachytherus mendocensis* from the late Eocene or early Oligocene Divisadero Largo Formation of Mendoza Province, Argentina (Shockey et al., 2007), but Cerdeño et al. (2006) have suggested that this specimen may actually derive from early Miocene rocks that overlie the Divisadero Largo Formation. The earliest secure records of the family come from the late Oligocene, when the family is represented by the genera *Anatrachytherus* and *Trachytherus* from Argentina and Bolivia (Reguero and Castro, 2004). The family reached its greatest diversity in the Miocene (Flynn et al., 2005), and mesotheriids persisted into the middle Pleistocene, in the form of the type genus, *Mesotherium* (McKenna and Bell, 1997). Mesotheriidae was one of only three notoungulate families to persist into the Quaternary, the others being Hegetotheriidae and Toxodontidae.

Classification

Within the order Notoungulata, Mesotheriidae is placed in the suborder Typotheria (Cifelli, 1993). In fact, Typotheria is named for the genus *Typotherium*, a synonym of *Mesotherium* (Shockey et al., 2007). In addition to Mesotheriidae, Typotheria traditionally includes other small bodied notoungulates in the families Oldfieldthomasiidae, Interatheriidae, and Archaeopithecidae (Simpson, 1967; McKenna and Bell, 1997). Recent opinion, however, favors inclusion of two additional families in Typotheria, Archaeohyracidae and Hegetotheriidae (Croft and Anaya, 2006). These families have traditionally been placed in a separate suborder, Hegetotheria, but phylogenetic studies indicate that their exclusion would render Typotheria paraphyletic (Cifelli, 1993; Billet et al., 2007). Within Typotheria, both Cifelli and Billet et al. indicate that mesotheriids are more closely related to archaeohyracids and hegetotheriids than to the remaining typotherian families. In fact, Billet et al.'s analysis indicates that both Mesotheriidae and Hegetotheriidae originated from within Archaeohyracidae.

McKenna and Bell (1997) recognized three subfamilies within Mesotheriidae: Fiandraiinae, Mesotheriinae, and Trachytheriinae. However, Flynn et al. (2005) have suggested that *Fiandraia*, the only known fiandraiine, is not a mesotheriid and may represent a toxodontid instead. Of the remaining subfamilies, Trachytheriinae includes earlier (Eocene and Oligocene) forms and may be paraphyletic with respect to Mesotheriinae, which includes more derived genera from the Miocene and later (Reguero and Castro, 2004).

Classification of Mesotheriidae:

- **Family †Mesotheriidae**

Source (edited): "http://en.wikipedia.org/wiki/Mesotheriidae"

Metailurus

Metailurus is a genus of false saber-toothed cat of the family Felidae, belonging to the tribe Metailurini endemic to North America, Africa, Europe, and Asia during the Miocene to Pleistocene, living from 9 Ma—11,000 years ago and existed for approximately 8.989 million years.

Metailurus was named by Zdansky (1924). It was assigned to Felidae by Carroll (1988).

Metailurus is known as a "false sabertooth" cat because, instead of true saberteeth, it has a cross between long, flat and blade-like saberteeth and the shorter cone-shaped teeth of modern-day felines. The canines are longer than even the Clouded Leopard but significantly shorter than true saber teeth and more conical than bladed. Due to the fragmentary nature of *Metailurus* fossils, the number of species is subject to debate.

Source (edited): "http://en.wikipedia.org/wiki/Metailurus"

Metamynodon

Metamynodon is an extinct genus of amynodont perissodactyls, and is among the longest lived genera of amynodonts, having first appeared during the late Eocene, and becoming extinct during the early Miocene, when it was supplanted by the semiaquatic rhinoceros, *Teleoceras*. Its fossils have been discovered in the United States (White River Fauna), Mongolia and China.

Characteristics

Restoration by Charles R. Knight

Restoration by Heinrich Harder

Metamynodon was about 4 metres (13 ft) in body length, and, although it was distantly related to the modern rhinoceros, looked more like a hippopotamus. Its front legs had four toes instead of the three found in modern rhinos. Although it was a herbivore (as indicated by its teeth), its skull had a bony ridge typically associated with carnivorous mammals. It probably fed on tough plant material, and the jaw muscles were attached to this ridge. *Metamynodon* would have used its enlarged canines to search for food in river banks, and may also have had highly flexible lips. Its eyes were placed high on the skull, meaning that it would have been able to see while almost fully submerged, much like a hippopotamus or crocodile.
Source (edited): "http://en.wikipedia.org/wiki/Metamynodon"

Metatomarctus

Metatomarctus is an extinct genus of Borophaginae and a terrestrial canine which inhabited most of North America during the Early Hemingfordian stage of the Miocene epoch living 23.0—5.3 Mya, existing for approximately 17.7 million years

Taxonomy

It was first described by Wang in 1999.

These carnivores hunted in packs, like modern canines, and may have preyed upon creatures such as *Equus*, rodents, and other smaller prey.

Morphology

Fossil specimens of two individuals' body mass were examined by Legendre and Roth. The first specimen was estimated to weigh 11.1 kg (24 lb). The second specimen was estimated to weigh 10.5 kg (23 lb).

Sister genera

Cormocyon, Desmocyon, Euoplocyon, Microtomarctus, Protomarctus, Psalidocyon, Tephrocyon.

Fossil distribution

- Thomas Farm Site, Gilchrist County, Florida, estimated age ~23.3—16.3 Mya.
- Pollack Farm Site, Kent County, Delaware, estimated age ~20.6—16.3 Mya.
- Hackberry Wash, San Bernardino County, California, estimated age ~20.6—16.3 Mya.
- Rattlesnake Hills, Fremont County, Wyoming, estimated age ~23.0—5.3 Mya.
- High Rock Canyon aka UCMP V-110, Humboldt County, Nevada estimated age ~23.0—5.3 Mya.
- Ahren's Prospect, Elder Ridge, and Pebble Creek, Runningwater Formation, Dawes County, Nebraska, estimated age ~20.6—16.3 Mya.

The genus contains one species: *M. canavus*. Other fossil locations: Maryland, Wyoming, New Mexico, and several in western Nebraska.
Source (edited): "http://en.wikipedia.org/wiki/Metatomarctus"

Metaxytherium

Metaxytherium is an extinct genus of dugong that lived from the Miocene to the Pleistocene. Its remains have been found in Africa, Asia, Europe, North America, and South America.

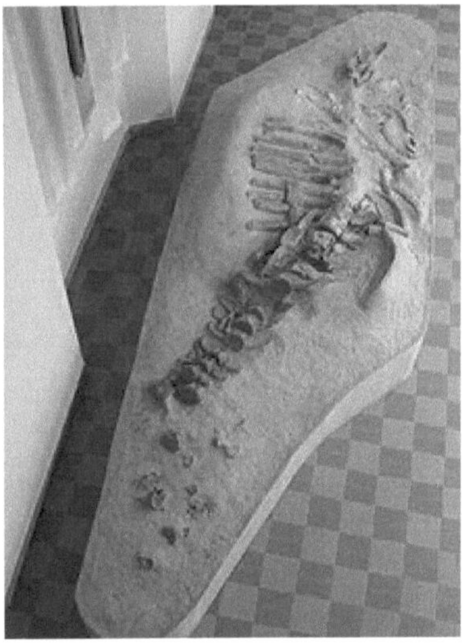

Metaxytherium subappenninum

Source (edited): "http://en.wikipedia.org/wiki/Metaxytherium"

Michenia

Michenia is an extinct genus of terrestrial herbivore the family Camelidae, endemic to North America from the Pliocene through Pleistocene 24.8—4.9 mya, existing for approximately 19.9 million years.

Taxonomy

Michenia was named by Frick and Taylor (1971). Its type is *Michenia agatensis*. It was assigned to Camelidae by Frick and Taylor (1971) and Carroll (1988).

Morphology

Two specimens were examined for estimated body mass by M. Mendoza, C. M. Janis, and P. Palmqvist. These specimens were estimated to weigh:

- 224.9 kg (500 lb)
- 121.9 kg (270 lb)

Fossil distribution

Fossil distribution ranged from California to Texas to Alberta, Canada and to Idaho and Nebraska.

Source (edited): "http://en.wikipedia.org/wiki/Michenia"

Migmacastor

Migmacastor is an extinct member of the beaver family, Castoridae, known from a single species, ***Migmacastor procumbodens***. Only a single specimen has been reported, a skull from the late Oligocene or early Miocene of Nebraska (Korth & Rybczynski 2003, p. 667). Features of the incisor teeth of *Migmacastor* indicate that they were used to dig. Other extinct beavers including the better-known *Palaeocastor* were also fossorial (digging), but Korth & Rybczynski (2003, p. 674) suggested that *Migmacastor* may have become a burrower independently.

Source (edited): "http://en.wikipedia.org/wiki/Migmacastor"

Miomachairodus

Miomachairodus was an extinct genus of large saber-toothed cats of the subfamily Machairodontinae, that lived in Europe, Asia, Africa, and North America during most of the Miocene. It survived until the Late Miocene (early Vallesian).

Miomachairodus coexisted with the barbourofelid *Barbourofelis* in Anatolia.

Source (edited): "http://en.wikipedia.org/wiki/Miomachairodus"

Mionothropus

Mionothropus is an extinct genus of nothrotheriine nothrotheriid sloth which existed in Peru in western Amazonia, during the late Miocene (Huayquerian age). It is known from the holotype LACM 4609/117533, which includes cranial remains, recovered from the riverbank deposits of the Río Acre region, were originally assigned to *Nothropus priscus*. It was first named by Gerardo De Iuliis, Timothy J. Gaudin and Matthew J. Vicars in 2011 and the type species is *Mionothropus cartellei*. The generic name is derived from "Mio", is a reference to the Miocene provenance of the holotype, and "nothropus" refers to the initial allocation of the holotype to this genus from the South American Pleistocene. The specific name honours Dr. Castor Cartelle.

Phylogeny

Cladogram after De Iuliis, Gaudin & Vicars, 2011:
Source (edited): "http://en.wikipedia.org/wiki/Mionothropus"

Miotapirus

Miotapirus harrisonensis is an extinct species of tapir lived during the early Miocene Epoch some 20 million years ago in North America.

Physically *Miotapirus* was virtually identical to its modern relatives; with a length of 2 m (6 ft 8 in) it was even the same size. Most likely it was also nocturnal and very adaptable.
Source (edited): "http://en.wikipedia.org/wiki/Miotapirus"

Miotylopus

Miotylopus is an extinct genus of terrestrial herbivore the family Camelidae, endemic to North America from the Late Oligocene through Early Miocene 24.8—20.6 mya, existing for approximately 4.2 million years.

Taxonomy

Miotylopus was named by Schlaikjer (1935). It is not extant. Its type is Miotylopus bathygnathus. It was assigned to Camelidae by Schlaikjer (1935) and Carroll (1988).

Morphology

Four specimens were examined for estimated body mass by M. Mendoza, C. M. Janis, and P. Palmqvist. These specimens were estimated to weigh:

- 77.0 kg (170 lb)
- 75.6 kg (170 lb)
- 59.7 kg (130 lb)
- 53.8 kg (120 lb)

Fossil distribution

Fossil distribution is restricted to Wyoming and Nebraska to Southern California.
Source (edited): "http://en.wikipedia.org/wiki/Miotylopus"

Moghradictis

Moghradictis is an extinct species of carnivorous cat-like mammals belonging to the superfamily Aeluroidea, endemic to North Africa from the Miocene 23.03—15.97 Ma, existing for approximately 7.06 million years.

Moghradictis is shown to have an omnivorous diet or more precisely, hypercarnivorous to hypercarnivorous.

Taxonomy

Moghradictis was named by Morlo et al. (2007). It is not extant. It was assigned to Stenoplesictinae by Morlo et al. (2007).
Source (edited): "http://en.wikipedia.org/wiki/Moghradictis"

Moropus

Moropus (meaning "slow foot") is an extinct genus of mammal, belonging to a group called chalicotheres, which were perissodactyl ("odd-toed") mammals, endemic to North America during the Miocene from living from ~23.03—5.3 Mya, existing for approximately 17.73 million years.

Moropus is related to the modern horse, rhino, and tapir.

Taxonomy

Moropus was named by Marsh (1877). Its type is ***Moropus distans***. It was synonymized subjectively with *Macrotherium* by Osborn (1893). It was assigned to Moropodidae by Marsh (1877); to Chalicotheriidae by Marsh (1877), Peterson (1907), Skinner (1968), Coombs (1978), Carroll (1988), Coombs (1998) and Holbrook (1999); and to Schizotheriinae by Geraads et al. (2007).

Morphology

Like other chalicotheres, they differed from their modern relatives in having large claws, rather than hooves, on the front feet; these claws may have been used for defense or digging for food. *Moropus* stood about 8 feet (2.4 m) tall at the shoulder. The three highly compressed claw-like hooves on each foot were split down the middle. These claws actually gave *Moropus* its name: " slow "or "sloth foot". This name implies that because of the claws, *Moropus* was a clumsy mover. But the articulation of the phalangeal (finger) bones, in addition to probable large foot and toe pads, shows that *Moropus* probably could raise the claws slightly to enable it to move about quite smoothly. Because the hooves curve inward, it probably had a pigeon-toed gait.

Body mass

Two specimens were examined by M. Mendoza, C. M. Janis, and P. Palmqvist for body mass.

- Specimen 1: 95.5 kg (210 lb)
- Specimen 2: 134.2 kg (300 lb)

Fossil distribution

- Phillips Ranch, Kern County, California estimated age: ~18.7 Mya.
- Stewart Spring(UCMP 2027), Mineral County and Esmeralda County, Nevada estimated age: ~18.7 Mya.
- Stage Hill I, aka Millennium's End Quarry, Scotts Bluff County, Nebraska estimated age: ~21.6—21.5 Mya.
- Sucker Creek site, Sucker Creak Formation, Malheur County, Oregon ~16.4 Mya.

Species

M. elatus

M. elatus specimen, AMNH

M. elatus was named by Marsh (1877).

Body mass

Two specimens were examined by M. Mendoza, C. M. Janis, and P. Palmqvist for body mass.

- Specimen 1: 118.4 kg (260 lb)
- Specimen 2: 296.8 kg (650 lb)

Fossil distribution

- Granby site, Grand County, Colorado ~23 Mya.
- Agate Springs Quarries, Sioux County, Nebraska, estimated age: ~20.9—20.8 Mya.
- American Museum-Cook Quarry, Sioux County, Nebraska, estimated age: ~23.03—5.33 Mya.
- Cart Trail Quarry, Box Butte County, Nebraska, estimated age: ~23.0—20.3 Mya.
- Morava Ranch Quarry, Box Butte County, Nebraska, estimated age: ~21.6 Mya.

M. hollandi

M. hollandi was named by Peterson (1907).

Fossil distribution

- Chugwater (aka Chugwater Water Hole), Platte County, Wyoming, estimated age: ~20.7 Mya.
- Jay Em, Goshen County, Wyoming, estimated age: ~20.7 Mya.
- Niobrara Canyon, Sioux County, Nebraska, estimated age: ~20.7 Mya.

M. matthewi

M. matthewi was named by Holland and Peterson, 1913-1914.

M. merriami

M. merrami was named by Holland and Peterson (1914). It was recombined as *Macrotherium merriami* by Matthew (1929) and Stirton (1939); it was recombined as *Chalicotherium merriami* by von Koenigswald (1932).

Fossil distribution

- High Rock Canyon, Humboldt County, Nevada, estimated age: ~17.2 Mya.
- Virgin Valley, Humboldt County, Nevada, estimated age: ~16.3 Mya.
- Humbug Quarry, Sioux County, Nebraska, ~16.5—16.25 Mya.
- Echo Quarry, Sioux County, Nebraska, ~16.3—13.6 Mya.

M. oregonensis

M. oregonsis was named by Leidy 1873. It was named by Leidy (1873) and recombined as *Moropus oregonensis* by Holland and Peterson (1914) and M. C. Coombs in 1978 and 1998, and also by M. C. Coombs, R. M. Hunt, E. Stepleton, L. B. Albright, III, and T. J. Fremd.

Body mass

Two specimens were examined by M. Mendoza, C. M. Janis, and P. Palmqvist for body mass.

- Specimen 1: 58.4 kg (130 lb)
- Specimen 2: 90.3 kg (200 lb)

Fossil distribution

- Johnson Canyon, John Day Formation, Wheeler County, Oregon, estimated age: ~22.2—21.9 Mya.
- Rose Creek, John Day Formation, Wheeler County, Oregon
- Toledo Bend, Fleming Formation, Newton County, Texas, estimated age: 21.9 Mya.
- St. Marks River, Leon County, Florida, estimated age: ~23.1—21.9 Ma.
- Buda Mine site, Alachua County, Florida, estimated age: ~23.1—23 Ma.

M. senex

M. sexex was named by Marsh (1877). It was considered a nomen dubium by Coombs (1978) and Coombs (1998).

Source (edited): "http://en.wikipedia.org/wiki/Moropus"

Morotopithecus

Morotopithecus bishopi is a species of fossil ape discovered in Moroto, Uganda.

Source (edited): "http://en.wikipedia.org/wiki/Morotopithecus"

Mylodontidae

Mylodontidae is a family of extinct mammals within the order of Pilosa and suborder Folivora living from approximately 23 mya—11,000 years ago, existing for approximately 22.89 million years. This family of ground sloths is related to the other families of extinct ground sloths, being the Megatheriidae, the Nothrotheriidae, the Orophodontidae and the Scelidotheriidae. The only extant families of the suborder Folivora are the Bradypodidae and the Megalonychidae. Phylogenetic analyses using homologous sequences from all extant edentate groups indicates that the Mylodontidae were closer related to Megalonychidae than to Bradypodidae.

The mylodontids together with their relatives the scelidotheriids and the orophodontids form the Mylodonta, the second radiation of ground sloths. The discovery of their fossils in caverns associated with human occupation lead some early researchers to theorize that the early humans built corrals when they could procure a young ground sloth, to raise the animal to butchering size. However, radiocarbon dates do not support simultaneous occupation of the site by humans and sloths. Subfossil remains like coproliths, fur and skin have been discovered in some quantities.

FAMILY †MYLODONTIDAE Gill, 1872

- *Urumacotherium* (*incertae sedis*)
- Subfamily Lestodontinae
  - Tribe Thinobadistini
    - Genus *Thinobadistes*
    - Genus *Sphenotherus*
  - Tribe Glossotheriini
    - Genus *Acremylodon*
    - Genus *Ranculcus*
    - Genus *Glossotherium*
    - Genus *Paramylodon*
    - Genus *Mylodonopsis*
  - Tribe Lestodontini
    - Genus *Lestodon*
    - Genus *Lestodontidion*
- Subfamily Mylodontinae
  - Genus *Glossotheriopsis*
  - Genus *Promylodon*
  - Genus *Strabsodon*
  - Genus *Megabradys*
  - Genus *Pleurolestodon*
  - Genus *Mylodon*

Source (edited): "http://en.wikipedia.org/wiki/Mylodontidae"

Nakalipithecus

Nakalipithecus nakayamai is a prehistoric great ape species that lived in today's Kenya region early in the Late Miocene, 10 million years ago (mya). It is the type species of the new genus *Nakalipithecus*. This ape was described from a fossil jawbone and eleven isolated teeth excavated in 2005 by a team of Japanese and Kenyan researchers in mud flow deposits in Nakali region of northern Kenya's Rift Valley Province, giving the genus its scientific name which means "Nakali ape".

The fossil teeth were covered in thick enamel, suggesting that the diet of this hominoid included a considerable amount of hard objects, possibly nuts or seeds. According to Kyoto University researchers, the *Nakalipithecus* species is very close to the last common ancestor of gorillas, chimpanzees, and humans. It can therefore be considered a basal member of the Homininae, before they split up into the three lineages alive today. *Nakalipithecus* also resembles the genus *Ouranopithecus*, another prehistoric hominid species found in present-day Greece.

The evolutionary importance of *Nakalipithecus* is twofold: first, together with *Ouranopithecus* it provides evidence that the Homininae lineages of today diverged no earlier than some 8 million years ago. Second, it supports the theory that the closest relatives of humans evolved in Africa. The competing view - that modern-type Great Apes went extinct in Africa and that the Homininae were originally an Asian lineage which only later recolonized Africa - is hard to reconcile with the early Late Miocene presence of a basal hominine in Africa.

The name of the genus refers to Nakali, the region where the fossil was found, while the species is named after Japanese geologist Katsuhiro Nakayama who died while working on the project.

Source (edited): "http://en.wikipedia.org/wiki/Nakalipithecus"

Nambaroo

Nambaroo gillespieae is an extinct genus of macropod marsupial from the late Oligocene some 25 million years ago of Australia.

Source (edited): "http://en.wikipedia.org/wiki/Nambaroo"

Nannippus

Nannippus is an extinct genus of horse endemic to North America during the Miocene through Pliocene, ~13.3—3.3 Ma, living approximately 11.1 million years.

Nannippus lived as far south as Central Mexico (*N. peninsulatus*) to as far north as Canada (*N. lenticularis*), to California in the west, North Carolina (*N. lenticularis*) and Florida (*N. peninsulatus*) in the east.

Species

- *N. aztecus* (Mooser 1968). Widespread in Florida and also found in Texas, Oklahoma, and Chihuahua, became extinct ~11.2—5.7 Ma.

- *N. beckensis* (Dalquest and Donovan, 1973) found in Texas only and became extinct ~3.4 Ma.
- *N. lenticularis* (Cope, 1893). found in Alberta, Canada, North Carolina, Alabama, Nebraska, and Kansas becoming extinct ~13 Ma.
- *N. montezumae* (Leidy, 1882).
- *N. morgani* (Hulbert, 1993) restricted to Florida. Appears to become extinct ~8.6 Ma.
- *N. peninsulatus* (Cope, 1893) found in Florida, Texas, Oklahoma, New Mexico, Arizona, and Mexico becoming extinct ~3.3 Ma.
- *N. parvulus* (Marsh, 1868).
- *N. westoni* (Simpson, 1930) restricted to Florida. Became extinct ~9.1—8.7 Ma.

Source (edited): "http://en.wikipedia.org/wiki/Nannippus"

Nanotragulus

Nanotragulus is an extinct genus belonging to the family Hypertragulidae, within the order Artiodactyla, endemic to North America, Europe, and Asia during the Eocene through Miocene, living 46.2—13.6 Ma, existing for approximately 32.6 million years.

Nanotragulus are primitive and ancient ruminants, resembling small deer or musk deer, although they were more closely related to modern chevrotains. It's diet is stated to be that of a frugivore.

Taxonomy

Nanotragulus was named by Lull (1922). It was assigned to Cervidae by Cook (1934); and to Hypertragulidae by Lull (1922), Matthew (1926) and Carroll (1988).

Morphology

Body mass

Four specimens were examined by M. Mendoza, C. M. Janis, and P. Palmqvist for body mass:

- 8.29 kg (18 lb)
- 3.84 kg (8.5 lb)
- 4.70 kg (10 lb)
- 4.54 kg (10 lb)

Fossil distribution

Partial list of fossil sites:

- Kimberly Member, Grant County, Oregon
- North Boulder Valley, Jefferson County, Montana
- Brooksville 2 Site, Hernando County, Florida

Source (edited): "http://en.wikipedia.org/wiki/Nanotragulus"

Naraborycytes

Naraborycytes is an extinct genus of marsupial mole from early Miocene deposits of Boodjamulla National Park of Riversleigh area, northwestern Queensland, Australia. It was first named by Michael Archer, Robin Beck, Miranda Gott, Suzanne Hand, Henk Godthelp and Karen Black in 2011 and the type species is *Naraboryctes philcreaseri*. The generic name means "to drink" (naraba in Garrawa and Waanyi languages of northwestern Queensland) in reference to its rainforest palaeohabitat + "digger" (oryctes in Greek) in reference to its fossorial specializations and close relationship to Notoryctes, and the specific name honors Phil Creaser.

Source (edited): "http://en.wikipedia.org/wiki/Naraboryctes"

Necrolestes

Necrolestes patagonensis ("Grave Robber" or "Thief of the Dead" + Patagonia) is an extinct mammal, suspected to be a marsupial, that lived in the early Miocene of South America.

About one-third of the animal's skeleton - including most of the skull - has been found as disassociated bones of several individuals. The jaw bends up at the tip, possibly supporting a fleshy appendage similar to the sensitive tentacles of the star-nosed mole. *Necrolestes* is also sometimes reconstructed as a mole-like creature. It probably fed on insects or worms.

Its classification is not firmly resolved due to it being highly apomorphic and having an anatomy unlike any other known mammal, living or extinct. It most likely belongs to the marsupial lineage (Metatheria). On the other hand, the possibility that it is an eutherian cannot be ruled out based on the currently available data, given as how, even as an island, South America had extensive lineages of both marsupial and placental mammals. A second species, *N. mirabilis* has now been described.

Source (edited): "http://en.wikipedia.org/wiki/Necrolestes"

Necromanis

Necromanis (literally "dead pangolin") is an extinct genus of pangolin from the Miocene of France. *Necromanis* is descended from the Eocene pangolins of genus *Eomanis*.

From what can be derived from the fossil specimens, the species of *Necromanis* were, for the most part, identical to modern *Manis* pangolins in anatomy, diet, and behavior.

There are at least three species recognized.

Source (edited): "http://en.wikipedia.org/wiki/Necromanis"

Neocynodesmus

Neocynodesmus is an extinct genus of terrestrial carnivore within the family Canidae which inhabited North America from the Oligocene to Early Miocene 30.8—-20.6 Mya existing for approximately 10.2 million years.

Taxonomy

Neocynodesmus was named by Macdonald (1963). It was assigned to Canidae by Macdonald (1963) and Carroll (1988).

Source (edited): "http://en.wikipedia.org/wiki/Neocynodesmus"

Neotherium

Neotherium mirum is an extinct species of walrus. It was smaller than living forms and it did not have long tusks. Males were larger than females.

Source (edited): "http://en.wikipedia.org/wiki/Neotherium"

Nesodon

Nesodon ("island tooth") is a genus of Miocene mammal belonging to the extinct order Notoungulata which inhabited southern South America during the Late Oligocene to Miocene living from 29.0—16.3 Ma and existed for approximately 12.7 million years.

Taxonomy

Nesodon imbricatus

Promegatherium nanum and *Nesodon*

Nesodon was named by Owen (1846). It was assigned to Toxodontidae by Carroll (1988). It was an early member of the family Toxodontidae, which included the well-known Pleistocene genus *Toxodon*. Like almost all toxodontids, *Nesodon* was endemic to South America. In particular, fossils of *Nesodon* are known from late early Miocene (Santacrucian SALMA) deposits of Argentina and Chile.

Three species of *Nesodon* are recognized including a larger species, *N. imbricatus*, and a smaller species, *N. conspurcatus*. A poorly-known and possibly invalid third species, *N. cornutus*, was similar to *N. imbricatus* but may have had a small horn on its head. All species of *Nesodon* were larger than species of the contemporary toxodontid *Adinotherium*.

The dentition of *Nesodon* shows features typical of living grazing (grass-eating) mammals, but a study of wear on the enamel of *N. imbricatus* suggests that it was a browser (leaf eater) that may have supplemented its diet with fruit or bark.

Source (edited): "http://en.wikipedia.org/wiki/Nesodon"

Nimbacinus dicksoni

Nimbacinus dicksoni was an ancient relative of the modern but extinct Thylacine. It lived approximately 23-16 million years ago in the Miocene period. *Nimbacinus dicksoni* was about 1.6 ft (50 cm) long. Being a predator, it probably ate birds, small mammals, and reptiles. Like the modern thylacine, it may have been an awkward runner and used stamina to catch prey rather than speed. Fossils have been found in Australia at Riversleigh in north-western Queensland and Bullock Creek in the Northern Territory. The fossils are very well preserved.

Source (edited): "http://en.wikipedia.org/wiki/Nimbacinus_dicksoni"

Nimbadon

Nimbadon is an extinct genus of Diprotodontia that lived in the Miocene Epoch. Many fossils have been found in the Riversleigh World Heritage propertyin north-western Queensland.

In 1990, skulls were unearthed in a previously unknown cave in the region. Researcher's estimate that Nimbadon first appeared about 15 million years ago and died out about 12 million years ago, perhaps from climate change-induced habitat loss.

Nimbadon lavarackorum

Nimbadon lavarackorum is described as being sheep-like. It is known from as many as 24 well-articulated specimens. The species was a browser, mainly feeding on stem and leaf litter on the forest floor. The feet and claws were large, being superficially similar to those of the Koala. They retracted their claws when walking.

Source (edited): "http://en.wikipedia.org/wiki/Nimbadon"

Nimravides

Nimravides was a prehistoric saber-toothed felid that lived in North America during the middle and late Miocene. Despite its scientific name *Nimravides* does not belong to the Nimravidae, but is a true felid, belonging to the family Felidae.

The earlier species of the genus, like *Nimravides pedionomus* from Chlaredonian deposits are smaller and more primitive than later forms, like *Nimravides catacopsis*. The tiger-sized *Nimravides catacopsis* strongly resembles the primitive *Machairodus aphanistus*, a species known from the Late Miocene of Eurasia .

Source (edited): "http://en.wikipedia.org/wiki/Nimravides"

Nothokemas

Nothokemas is an extinct genus of terrestrial herbivore the family Camelidae, endemic to North America from the Oligocene through Miocene 30.8—20.430 mya, existing for approximately 10.757 million years.

Taxonomy

Nothokemas was named by White (1940) [also said to be 1947]. It was assigned to Camelidae by White (1940) and Honey et al.

Morphology

Four specimens were examined for estimated body mass by M. Mendoza, C. M. Janis, and P. Palmqvist. These specimens were estimated to weigh:

- 73.8 kg (160 lb)
- 78.8 kg (170 lb)
- 48.0 kg (110 lb)
- 52.3 kg (120 lb)

Fossil distribution

Fossil distribution is concentrated along the Gulf Coast from Texas to Florida in 11 different sites.

Source (edited): "http://en.wikipedia.org/wiki/Nothokemas"

Notiomastodon

Notiomastodon is an extinct genus of proboscidea endemic to South America during the Pleistocene from 1.810 Ma—780,000 years ago, living for approximately 1.019 million years.

It had two tusks on either side of a trunk like other members of Gomphotheriidae. Its apparent range was the central to southern continent with fossils recovered in Tarija Province, Bolivia. It would have probably shared its habitat with Haplomastodon based on genus age of existence and geography.

Taxonomy

Notiomastodon was named by Cabrera (1929). It is not extant. It was assigned to Gomphotheriidae by Carroll (1988).

Source (edited): "http://en.wikipedia.org/wiki/Notiomastodon"

Notoungulata

Notoungulata is an extinct order of hoofed, sometimes heavy bodied mammalian ungulates which inhabited South America during the Paleocene to Pleistocene, living from approximately 57 Ma to 11,000 years ago.

Taxonomy

Notoungulata was named by Roth (1903). It was assigned to Eutheria by Carroll (1988). The order includes the huge *Toxodon*. Due to the isolated nature of South America, many notoungulates evolved along convergent lines into forms that resembled mammals on other continents. Examples of this are *Pachyrukhos*, a notoungulate that evolved to fill the role of rabbits and hares and *Homalodotherium*, which evolved to resemble chalicotheres. During the Pleistocene, *Toxodon* was the largest common notoungulate. Most of the group (*Toxodon* being an exception) became extinct after the landbridge between North and South America was formed, allowing North American ungulates to enter South America in the Great American Interchange and outcompete the native fauna. This order is united with other South-American ungulates in the super-order Meridiungulata.

Cifelli has argued that Notioprogonia is paraphyletic, as it would include the ancestors of the remaining suborders. Similarly, Cifelli indicated that Typotheria would be paraphyletic if it excluded Hegetotheria and he advocated inclusion of Archaeohyracidae and Hegetotheriidae in Typotheria.

As reflected in the map that accompanies this article, Notoungulata were for many years taken to include the order Arctostylopida, whose fossils are found mainly in China. Recent studies, however, have concluded that Arctostylopida are more properly classified as gliriforms.

Orders and families

- Order Notoungulata - notoungulates
  - Suborder Notioprogonia
    - Family Henricosborniidae
    - Family Notostylopidae
  - Suborder Toxodonta
    - Family Isotemnidae
    - Family Leontiniidae
    - Family Notohippidae
    - Family Toxodontidae
    - Family Homalodotheriidae
  - Suborder Typotheria
    - Family Archaeopithecidae
    - Family Oldfieldthomasiidae
    - Family Interatheriidae
    - Family Campanorcidae
    - Family Mesotheriidae
  - Suborder Hegetotheria
    - Family Archaeohyracidae
    - Family Hegetotheriidae

Source (edited): "http://en.wikipedia.org/wiki/Notoungulata"

Nuralagus rex

Restoration of *Nuralagus rex*

The **Minorcan Giant Lagomorph**, *Nuralagus rex*, is an extinct rabbit that lived in the island of Minorca from the Messinian until around the middle of the Pliocene, when it became extinct (3 to 5 million years ago). During that time, both Majorca and Minorca were united in one large landform, allowing *Myotragus balearicus* to colonize the Minorcan Giant Lagomorph's habitat.

Fossils have been found of individuals which could have weighed up to 23 kg (50 lbs), although the estimated average body mass of the species is about 12 kg. They had small ears, tiny eyes, short legs and a short, straight and rigid vertebral column, very different from their living relatives.

According to paleontologist Josep Quintana Cardona, *N. rex* didn't hop much if at all, as evident by its relatively short, stiff spine. No fossil remains of any rabbit eating predators have been found, so it is suggested that the lack of predators meant they didn't have a reason to maintain speed and agility, and could evolve to larger sizes.

Source (edited): "http://en.wikipedia.org/wiki/Nuralagus_rex"

Obdurodon

Obdurodon is an extinct monotreme genus containing three species. *Obdurodon* differed from modern Platypuses in that it had molar teeth (in the modern platypus, individuals lose their teeth upon reaching adulthood).

Species

Obdurodon dicksoni

- Discovered in 1984 by Michael Archer, F. A. Jenkins, S. J. Hand, P. Murray, and H. Godthelp, at Riversleigh in North West Queensland.
- Habitat : Queensland
- Epoch : Lower and middle Miocene
- This species is characterized by a skull and several scattered teeth. Physically, it looked much like modern Platypus although there are a few significant differences.
- The holotype is kept at the Queensland Museum in Brisbane.
- *Obdurodon dicksoni* is bigger than a Platypus.
- The septomaxilla (a part of the upper jawbone) of *O. dicksoni* is bigger than for the Platypus, which supposes a hypertrophied beak.
- The coronoid process and angulary process of *O. dicksoni* have quite disappeared in the Platypus, leaving the Platypus's skull flat on the sides. This indicates that the mastication technique of *O. dicksoni* was different from that of the Platypus, using the muscles anchored to these processes.
- *Obdurodon dicksoni*'s beak has an oval hole surrounded by bones in the center whereas the Platypus' beak has a V-shape and isn't stopped on top.
- *O. dicksoni'* has molars, whereas the Platypus uses keratinized pads (only young Platypuses have teeth).

- It seems, due to the shape of the beak, that *O. dicksoni* nourished itself digging in the sides of rivers whereas the Platypus digs in the bottom of the river.

O. dicksoni' has (like the Platypus) shearing crests instead of incisor and canine teeth. It bore two premolars and three molars on each side of the lower jaw. The M1 had six roots, the M2 had five, and the M3 only one. The upper jaw bore two premolars and two molars on each side. The M1 had six roots, the M2 four. The premolars had only one root and a very different shape from the molars. They were separated from the shearing crests by an area without dentition. The roots of the molars were barely a third as high as the crown. Molars had only been found apart from skulls, implying that they weren't well-anchored.

Obdurodon insignis

- Discovered in 1975 by Mike O. Woodburne and Dick H. Tedford at Etudunna Formation in the desert of Tirari.
- Habitat : South Australia
- Epoch : Upper Oligocene
- The holotype is an inferior left molar and is kept in the South Australia's Museum, Adelaide. The tooth has six roots. There also have been found M2 with four roots and fragments of jawbone and pelvis. Obdurodon insignis had one more canine tooth (NC1) than its ancestor *Steropodon galmani*. Its beak must have been proportionally smaller than the one of *Obdurodon dicksoni*.

Monotrematum sudamericanum

- Discovered in 1992 by Rosendo Pascual, Michael Archer, E. O. Juareguizar, J. L. Prado, H. Godthelp, and S. J. Hand, at Punta Peligro, Argentina.
- Habitat : Patagonia
- Epoch : Lower Paleocene (61 million years)
- *Monotrematum sudamericanum* is now more often held as part of the same genus as *Obdurodon*. It is known only from two lower and one upper platypus teeth. It is the only known non-Australasian ornithorhynchid. The main difference, apart from continent and age, is its size: the teeth of *Monotrematum* are around twice as large as other similar species. These fossils presently reside in the collections of Museo de La Plata and Museo Paleontológico Egidio Feruglio, both in Argentina.
- According to Pascual, "The preserved enamel in the central region shows that the crown pattern is almost identical to that of *Obdurodon*: it is composed of two V-shaped lobes, the anterior of which is wider, separated from the posterior one by a valley that connects the lingual and buccal sides of the crown separating the anterior and posterior lobes."

Source (edited): "http://en.wikipedia.org/wiki/Obdurodon"

Oioceros

Oioceros is an extinct genus of sheep from the late Miocene. Its fossils have been found in Greece, China, and Africa. It was first discovered by Wagner in 1857, and contains nine species, *O. rothii*, *O. atropatenes*, *O. jiulongkouensis*, *O. noverca*, *O. robustus*, *O. stenocephalus*, *O. lishanensis*, *O. wegneri*, and *O. tanyceras*. Former species include *O. grangeri* (Pilgrim, 1934), now recognized as the genus *Sinomegoceros*, and *O. xiejiaensis* (Li and Qui; 1980), now recognized as the genus *Sinopalaeoceros*.

Oioceros was somewhat resemblant of a gazelle, and represents the world's oldest-known bovid.

Description

In 1997, an almost complete skull of *O. rothii* proved that it was similar to *Samotragus praecursor*, with a short face, compared to its relatively long braincase, and isolated horn cores. However, unlike *S. praecursor*, *O. rothii* had a jaw of primarily premolar teeth, which were longer than the molars found in the latter. Pilgrim and Hopwood described the genus as being "small size with long slender muzzle; face bent down on basicranial axis either slightly or to a moderate extent; orbits with expanded orbital roof; horn-cores twisted counter-clockwise in a fairly close spiral of one or two revolutions, widely separate, tilted backward or fairly upright, divergent, with a cross-section almost circular or elliptical, keeled either anteriorly or posteriorly or both; dentition moderately hypsodont, premolar series rather long and slender, molars broad with ribs of medium strength" (Pilgrim and Hopwood, 1928, p. 24). The horn cores of *Oioceros* have a torsion.

Species

- ***Oioceros rothii*** was first named by Wagner in 1857. He originally had the specimen he had found in Pikermi, Greece, under the name *"Antilope rothii"*. In 1902, Gaillard reassigned the species to *Oioceros rothii*.
- ***Oioceros atropatenes*** was first named by Rodler & Weithofer in 1890.
- ***Oioceros jiulongkouensis*** was first named by G. Chen and W. Wu in 1976.
- ***Oioceros noverca*** was first named by Pilgrim in 1934.
- ***Oioceros robustus*** was first named by Chen and Wu in 1976.
- ***Oioceros stenocephalus*** was first named by Chen and Wu in 1976.
- ***Oioceros wegneri*** was first named by Andree in 1926. It possibly represents an entirely different genus of sheep.

Source (edited): "http://en.wikipedia.org/wiki/Oioceros"

Omoiosicista

Omoiosicista is an extinct genus of dipodid rodent which existed in central Nei Mongol, China, during the early Miocene (middle Burdigalian age). It was first named by Yuri Kimura in 2010 and the type species is *Omoiosicista fui*.

Source (edited): "http://en.wikipedia.org/wiki/Omoiosicista"

Ontocetus

Ontocetus is an extinct genus of walrus, an aquatic carnivoran of the family *Odobenidae*, endemic to coastal regions of the southern North Sea and the south-eastern coastal regions of the U.S. during the Miocene-Pleistocene. It lived from 13.6 mya—300,000 years ago, existing for approximately 13.3 million years.

Taxonomy

Originally named ***Alachtherium*** by Du Bus (1867) (type *Alachtherium cretsii.*). It was It was assigned to *Odobenidae* by Palmer (1904) and Kellogg (1922). *Ontocetus* was named by Joseph Leidy (1859). Its type is *Ontocetus emmonsi*. It was considered monophyletic by Uhen (2010). It was synonymized subjectively with *Hoplocetus* by Trouessart (1898); it was considered a *nomen dubium* by Fordyce and de Muizon (2001). It was assigned to *Sirenia* by Cope (1875); to *Physeterinae* by Hay (1902) and Simpson (1945); to *Physeteridae* by Gill (1871), Palmer (1904), Barnes (1977) and Carroll (1988); to *Cetacea* by Sepkoski (2002); to *Odobeninae* by Kohno (2006); to *Odobenini* by Kohno and Ray (2008); and to *Odobenidae* by Ray (1975) and Uhen (2010). *Ontocetus oxymycterus* was named by Kellogg (1925). Its type specimen is USNM 10923 and is a 3D body fossil. Its type locality is Santa Barbara Lighthouse, which is in a Serravallian marine dolomite in the Monterey Formation of California. It was recombined as *Scaldicetus oxymycterus* by Kohno and Ray (2008).

Source (edited): "http://en.wikipedia.org/wiki/Ontocetus"

Oreodontoides

Oreodontoides is an extinct genus of oreodont of the family *Merycoidodontidae*, subfamily *Merycoidodontinae*, endemic to North America during the Late Oligocene-Early Miocene subepochs (24.8—20.6 mya), existing for approximately 4.2 million years.

Taxonomy

Oreodontoides was named by Thorpe (1923) (named as subgenus of *Eporeodon* by Thorpe in 1921) and raised to genus level. It was synonymized subjectively with *Merychyus* by Loomis in 1924 and assigned to *Merycoidodontidae* by Thorpe in 1923 and 1937, (1923), by Schultz and Falkenbach (1947) and Lander (1998).

Morphology

A single specimen was examined by M. Mendoza for body mass and was estimated to have a weight of 34.9 kg (76.9 lbs).

Fossil distribution

Fossils have been uncovered throughout the western U.S. including Wheeler and Grant County, Oregon, Laramie County, Wyoming, Washabaugh County, South Dakota.

Species

Oreodontoides oregonensis (synonymous with *Desmatochoerus anthyonyi, Merychyus curtus*).

Source (edited): "http://en.wikipedia.org/wiki/Oreodontoides"

Oreopithecus

Oreopithecus is an extinct primate from the Miocene epoch whose fossils have been found in today's Tuscany and Sardinia in Italy. *Oreopithecus* existed 9 to 7 million years ago in the Tusco-Sardinian area when this territory was an isolated island in an arc of islands stretching from central Europe to northern Africa. It was one of a large number of European immigrants that settled this area in the Vallesian-Turolian transition and one of few hominoids to survive the so called Vallesian Crisis together with *Sivapithecus* in Asia. To date, dozens of individuals have been discovered at the Tuscan localities of Montebamboli, Montemassi, Casteani, Ribolla, and, most notably, in the fossil-rich lignite mine in the Baccinello Basin, making it one of the best-represented fossil apes.

Etymology

The scientific name *Oreopithecus* quickly inspired the commonly used though unofficial name for the species among paleontologists, Cookie Monster. The true etymology, however, is rather more mundane: it comes from the Greek ὄρος, oros and πίθηκος, pithekos, meaning "hill-ape".

Evolutionary history

Oreopithecus bambolii was first described by French paleontologist Paul Gervais in 1872. In the 1950s, Swiss paleontologist Johannes Hürzeler discovered a complete skeleton in Baccinello and claimed it was a true hominid — based on its short jaws and reduced canines, at the time considered diagnostic of the hominid family — and a biped — since the short pelvis was closer to those of hominids than those of chimpanzees and gorillas. However, *Oreopithecus'* hominid affinities remained controversial for decades until new analyses in the 1990s reasserted *Oreopithecus* as directly related to *Dryopithecus*; the peculiar cranial and dental features explained as consequences of insular isolation. These new evidences confirmed that *Oreopithecus* was bipedal but also revealed that its peculiar form of bipedalism was much different from that of *Australopithecus'* — the hallux formed a 100° angle with the other toes enabling the foot to act as a tripod in erect postures — and could not enable *Oreopithecus* to develop a fast bipedal locomotion. When a land bridge finally broke the isolation of the Tusco-Sardinian area 6.5 million years ago, truly large predators such as *Machairodus* and *Metailurus* were present among the new generation of European immigrants and *Oreopithecus* faced quick extinction together with other endemic genera.

Taxonomic classification

Known as the "enigmatic hominoid", *Oreopithecus* can dramatically rewrite the palaeontological map depending on if it is a descendant from the European ape *Dryopithecus* or some African anthropoid. Some have suggested the unique locomotory behavior of *Oreopithecus* requires a revision of the current consensus on the timing of bipedality in human developmental history, but there is limited agreement on this point among paleontologists.

Some researchers have related *Oreopithecus* to the early Oligocene *Apidium*, a small arboreal proto-ape that lived nearly 34 million years ago in Egypt. It shows strong links to modern apes in its postcranium and, in this respect, it is the most modern Miocene ape below the neck with closest similarities to the postcranial elements of *Dryopithecus*, but its dentition is adapted to a leafy diet and a close link is uncertain. Others claim it to be either the sister taxon to *Cercopithecoidea* or a even direct human ancestor, but it is usually placed in its own subfamily within *Hominidae*. It could instead be added to the same subfamily as *Dryopithecus*, perhaps as a distinct tribe (*Oreopithecinae*).

Physical characteristics

Oreopithecus bambolii is estimated to have weighed 30–35 kg (66–77 lb). It possessed a relatively short snout, elevated nasal bones, small and globular neurocranium, vertical orbital plane, and gracile facial bones. The shearing crests on its molars suggest a diet specializing in plant leaves. The very robust lower face, with a large attachment surface for the masseter muscle and a sagittal crest for attachment of the temporal muscle, indicates a heavy masticatory apparatus. Its teeth were small relative to body size. The lack of a diastema (gap) between the second incisor and first premolar of the mandible indicates that *Oreopithecus* had canines of size comparable to the rest of its dentition. In many primates, small canines correlate with reduced inter-male competition for access to mates and less sexual dimorphism.

Positional behavior

Its habitat appears to have been swampy, and not savanna or forest. The postcranial anatomy of *Oreopithecus* features adaptations to both suspensory arborealism and bipedalism. Functional traits related to suspensory locomotion include its broad thorax, short trunk, high intermembral index, long and slender digits, and extensive mobility in virtually all joints. At the same time, it also features adaptations to upright walking such as the presence of a lumbar curve, in distinction to otherwise similar species known from the same period. Since the fossils have been dated to about 8 million years ago, this represents an unusually early appearance of upright posture. How adapted it was for bipedal walking is not known, but its fingers and arms also seems to show adaptations for climbing and swinging.

Morphological and functional studies indicate a significant part of *Oreopithecus* positional behavior was bipedal. Its foot has been described as chimp-like, but is different from those of extant primates. The habitual line of leverage of the primate foot is parallel to the third metatarsal bone. In *Oreopithecus*, the lateral metatarsals are permanently abducted so that this line falls between the first and second metatarsals instead. Furthermore, the shape of the tarsus indicate loads on the foot were transmitted to the medial side of the foot instead of the lateral, like in other primates. The metatarsals are short and straight, but their lateral orientation increased the area of support during bipedal standing and walking.

The lack of predators and the limitation of space and resources in *Oreopithecus'* insular environment favored a locomotor system optimized for low energy expenditure rather than speed and mobility. Its peculiar feet and short legs were optimized for stable postural harvesting and bipedal shuffling rather than walking or running. The postcranial anatomy of *Oreopithecus* is therefore considered autapomorphic rather than synapomorphic. Its foot proportions are close to the unusual proportions of *Gorilla* and *Homo* but are distinct from those found in specialized climbers. It is not comparable to *Gorilla* due to low body mass but *is* similar to *Homo*; it is not our ancestor but a key species for understanding human bipedality.

Semicircular canals

The semicircular canals of the inner ear serves as a sense organ for balance and controls the reflex for gaze stabilization. The inner ear has three canals on each side of the head, and each of the six canals encloses a membranous duct that forms an endolymph-filled circuit. Hair cells in the duct's auditory ampulla pick up endolymph disturbances caused by movement, which register as rotatory head movement. They respond to body sway of frequencies greater than 0.1 Hz and trigger the vestibulocollic (neck) reflex and vestibuloocular (eye) reflex to recover balance and gaze stability. The bony semicircular canals allow estimates of duct arc length and orientation with respect to the sagittal plane.

During upright walking, human head rotation frequencies range between 2–8 Hz along the sagittal and coronal planes and between 1–3 Hz along the horizontal plane. Across species, the semicircular canals of agile animals have larger arcs than those of slower ones. For example, the rapid leaper *Tarsius bancanus* has semicircular canals much bigger than the slow-climbing *Nycticebus coucang*. The semicircular canals of brachiating gibbons are bigger than those of arboreal and terrestrial quadrupedal great apes. As a rule of thumb, arc size of the ducts *decreases* with body mass and consequently slower angular head motions. Arc size *increases* with greater agility and thus more rapid head motions. Modern humans have bigger arcs on their anterior and posterior canals, which reflect greater angular motion along the sagittal plane. The lateral canal has a smaller arc size, corresponding to reduced head movement from side to side.

Allometric measurements on the bony labyrinth of BAC-208, a fragmentary cranium that preserves a complete, undeformed petrosal bone suggest that *Oreopithecus* moved with agility comparable to extant great apes. Its anterior and lateral semicircular canal sizes fall within the range for great apes. Its relatively large posterior arc implies that *Oreopithecus* was more proficient at stabilizing angular head motion along the sagittal plane.

Dexterity

Oreopithecus had hominid-like hand proportions that allowed a firm, pad-to-pad precision grip. Features not present in the hands of extant or fossil apes include hand length, relative thumb length, a deep and large insertion for the flexor pollicis longus, and the shape of the carpometacarpal joint between the metacarpal bone of the index finger and the capitate bone. At the base of the second metacarpal bone, the facet for the capitate is oriented transversally, like in hominids. The capitate, on the other hand, lacks the waisting associated with apes and climbing, and still present in *Australopithecus*. *Oreopithecus* share the specialised orientation at the carpometacarpal joint with *A. afarenis* and the marked groove for the flexor pollicis longus with *A. africanus*. It is thus likely that the hand morphology of *Oreopithecus* is derived for apes and convergent for early hominids.

Source (edited): "http://en.wikipedia.org/wiki/Oreopithecus"

Oriomeryx

Oriomeryx is an extinct genus of Artiodactyla, of the family Moschidae (deer-like), endemic to Europe from the early Miocene epoch, 22.4—20.0 Ma, existing for approximately 2.4 million years.

Taxonomy

Oriomeryx was named by Ginsburg (1985). It was assigned to Pecora by Ginsburg et al. (1994); and to Moschidae by Prothero (2007).

Fossil distribution

- Cetina de Aragon, Zaragoza, Spain

Source (edited): "http://en.wikipedia.org/wiki/Oriomeryx"

Orophodontidae

Orophodontidae is a family of extinct ground sloths within the order of *Pilosa* and suborder *Folivora*. The name is often **disused** with genus members reassigned.

This family is related to the other families of extinct ground sloths, being the *Megatheriidae*, the *Mylodontidae*, the *Nothrotheriidae*, and the *Scelidotheriidae*. The only extant families of the suborder *Folivora* are the *Bradypodidae* and the *Megalonychidae*.

The orophodontids constitute a rather small but quite distinct group. They are classified as a distinct mylodontan superfamily **Orophodontoidea**, the sister taxon to the Mylodontoidea.

'*†Orophodontidae* members prior to reassignment.

- Genus *Proplatyarthrus, (Megalonychidae)*
- Genus *Orophodon, (Mylodontidae)*
- Genus *Octodontotherium, (Mylodontidae)*
- Genus *Octomylodon, (Mylodontidae)*
- Genus *Octodontobradys*

Source (edited): "http://en.wikipedia.org/wiki/Orophodontidae"

Orrorin

Orrorin tugenensis is considered to be the second-oldest (after *Sahelanthropus*) known hominin ancestor that is possibly related to modern humans, and it is the only species classified in genus ***Orrorin***. *Orrorin* is significant because it can be an early bipedal hominin.

The name was given by the discoverers who found *Orrorin* fossils in the Tugen Hills of Kenya. By analysing radiometric decay (K–Ar dating), paleomagnetism, and biochronology the age of the specimen have been estimated to 6 to 5.8 million years (Ma) (average: 6. 1 Ma). At present, 20 fossils have been found at four sites in the Lukeino Formation: of these, the fossils at Cheboit and Aragai are the oldest (6.1 Ma), while those in Kapsomin and Kapcheberek are found in the upper levels of the formation (5.7 Ma).

Fossils

The 20 specimen found this far include: the posterior part of mandible in two pieces; a symphysis and several isolated teeth; three fragments of femurs; a partial humerus; a proximal phalanx; and a distal thumb phalanx.

Orrorin had small teeth relative its body size. Its dentition differs from that found in *Australopithecus'* in that its cheek teeth are smaller and less elongated mesiodistally; and from *Ardipithecus* in that its enamel is thicker. The dentition differs from both these species in the presence of mesial groove on the upper canines. The canines are ape-like but reduced, like those found in Miocene apes and female chimpanzees. *Orrorin* had small post-canines and was microdont like modern humans, whereas *Australopithecus* was megadont.

In the femur, the head is spherical and rotated anteriorly; the neck is elongated and oval in section; and the lesser trochanter protrudes medially. While this suggest that *Orrorin* was bipedal, the rest of the postcranium indicates it climbed trees. while the proximal phalanx is curved, the distal pollical phalanx is of human proportions and have thus been associated with toolmaking, but should probably be associated with grasping abilities useful for tree-climbing in this context.

After the fossils were found in 2000, they were held at the Kipsaraman village community museum, but the museum was subsequently closed. Since then, according to the Community Museums of Kenya chairman Eustace Kitonga, the fossils are stored at a secret bank vault in Nairobi.

Classification

If *Orrorin* proves to be a direct human ancestor, then australopithecines such as *Australopithecus afarensis* ("Lucy") may be considered a side branch of the hominid family tree: *Orrorin* is both earlier, by almost 3 million years, and more similar to modern humans than is *A. afarensis*. The main similarity is that the Orrorin femur is morphologically closer to that of *H. sapiens* than is *Lucy's*; there is, however, some debate over this point.

Other fossils (leaves and many mammals) found in the Lukeino Formation show that *Orrorin* lived in dry evergreen forest environment, not the savanna assumed by many theories of human evolution.

Discovery

The team that found these fossils in 2000 was led by Brigitte Senut and Martin Pickford from the Muséum national d'histoire naturelle. The discoverers conclude that *Orrorin* is a hominin on the basis of its bipedal locomotion and dental anatomy; based on this, they date the split between hominins and African great apes to at least 7 million years ago, in the Messinian. This date is markedly different from those derived using the molecular clock approach, but has found general acceptance among paleoanthropologists.

Source (edited): "http://en.wikipedia.org/wiki/Orrorin"

Orycterocetus

Orycterocetus is an extinct genus of sperm whale from the Miocene.

Source (edited): "http://en.wikipedia.org/wiki/Orycterocetus"

Osbornoceros

Osbornoceros is a prehistoric antilocaprid genus belonging to the family Antilocapridae, all of whose species are extinct except for the pronghorn. *Osbornoceros osborni* is the only known species of the *Osbornoceros* genus. *Osbornoceros* lived during the Late Miocene around 7-6 million years ago in what is currently North America. It is well represented in fossil discoveries, with nearly a dozen specimens having been found to date. All come from the Chamita Formation in a quarry near Lyden, New Mexico, the site of numerous other finds such as that of *Chamitataxus*, a prehistoric badger who lived at the same time. The holotype specimen of *Osbornoceros* was discovered in 1937 and many more were found nearby during further expeditions.

Osbornoceros was strikingly similar to today's pronghorn; it was lightly built and had a series of small horns that protruded from its skull. It was, like its relatives, a quadruped herbivore and grazed on the grassy plains of its time. It is not known if *Osbornoceros* had any predators and much is still unknown about its paleobiology, but it is assumed that it was similar to the present day pronghorn and its extinct relatives. *Osbornoceros* was covered in a short fur and was most likely a good runner.

Source (edited): "http://en.wikipedia.org/wiki/Osbornoceros"

Osbornodon

Osbornodon ("Osborn's dog") is an extinct genus of bone-crushing canid (canidae) that were endemic to North America and which lived from the Orellan age of the Early Oligocene to Early Miocene epoch 33.9—15.97 Ma (AEO) and existed for approximately 17.93 million years.

Taxonomy

Osbornodon was named by Wang (1994). Its type is Osbornodon fricki. It was assigned to Canidae by Wang (1994) and Munthe (1998).

Morphology

Two fossil specimens were examined by Legendre and Roth for body mass. The first specimen was estimated to have weighed 22.2 kg (49 lb). The second specimen was estimated to have weighed 20.2 kg (45 lb).

Genus

Seven known species of the genus *Osbornodon* existed:

- *Osbornodon fricki* (18 Ma)
- *Osbornodon iamonensis* (synonyms: *Cynodesmus nobilis*, *Paradaphoenus tropicalis*) (21 Ma), *Osbornodon renjiei* (33 Ma), and *Osbornodon sesnoni* (32 Ma).

Species

Osbornodon scitulus appears to be a transitional species that partially fills a large morphological and stratigraphical gap within the *Osbornodon* genus. Compared to *Osbornodon renjiei* and *Osbornodon sesnoni*, *Osbornodon scitulus* possesses derived characters such as large frontal sinus, high sagittal crest, narrow infraorbital canal, short bulla, and broad premolars. *O. scitulus* differs from *Osbornodon wangi* in larger size and relatively wider P3 and shorter P4. *O. scitulus* is distinguishable from *Osbornodon iamonensis* and later species in its primitive characters such as a non-elongated rostrum, paroccipital process that is not expanded posteriorly, mastoid process not reduced, and p4 not differentially enlarged relative to p3 (Wang, 2003).

Source (edited): "http://en.wikipedia.org/wiki/Osbornodon"

Osbornodon iamonensis

Osbornodon iamonensis is an extinct species of bone-crushing Hesperocyoninae, a predecessor of modern dogs that were endemic to North America and which lived from the Orellan age of the Early Oligocene to Early Miocene epoch 23.6—16.3 Ma and existed for approximately 7.3 million years.

O. iamonensis was named by E. H. Sellards in 1916 for Lake Iamonia in northern Florida.

Taxonomy

Osbornodon iamonensis was named by E. H. Sellards (1916). Its type is FGS 5082. It was recombined as Cynodesmus iamonensis by Simpson (1932), Olsen (1956) and Wilson (1960); it was recombined as ***Osbornodon iamonensis*** by Xiaoming Wang (1994), X. Wang and Richard H. Tedford (1996) and Hayes (2000). It was assigned to Mesocyon by Sellards (1916); to Cynodesmus by Simpson (1932), Olsen (1956) and Wilson (1960); and to Osbornodon by Wang (1994), Wang and Tedford (1996) and Hayes (2000).

Morphology

Two fossil specimens were examined by Legendre and Roth for body mass.

- Specimen 1 was estimated to have weighed 17.6 kg (39 lb).
- Specimen 2 was estimated to have weighed 16.2 kg (36 lb).

Fossil distribution

- Griscom Plantation Site, northeast Leon County, Florida. ~23.6—18.8 Mya.
- Thomas Farm site in Gilchrist County, Florida. ~21.7 Mya.
- The remaining two specimens were found in the Dry Creek Prospect B Site and Bridgeport Quarries Site of Nebraska. ~20.6—16.3 Mya.

Source (edited): "http://en.wikipedia.org/wiki/Osbornodon_iamonensis"

Ouranopithecus

Ouranopithecus macedoniensis, sometimes called ***Graecopithecus freybergi***, is a prehistoric hominid species found in Greece and dated to the late Miocene. Based on *O. macedoniensis*'s dental and facial anatomy, it is possible that *O. macedoniensis* was a dryopithecine. However, *O. macedoniensis* seems to be more closely related to orangutans in subfamily Ponginae while the most of Dryopithecinae are more closely related to the other great apes in Homininae and a few are considered to be outside of the ape clade altogether. One distinctive trait that *Ouranopithecus* shares with the humans and other modern African apes is the *frontal sinus*, a cavity in the forehead. Some investigators consider it possible that *O. macedoniensis* was the last common ancestor of the great apes and the humans.

Morphology

O. macedoniensis had a large, broad face with a prominent supraorbital torus. It also had square-shaped orbits. *O. macedoniensis* may have had a relative large body size. *O. macedoniensis*'s molar enamel cover was fairly thick and had low cusps. Sexual dimorphism is evident by the teeth. The male *O. macedoniensis* had large canine teeth with shearing lower premolars.

Diet

Based on the heavily pitted surface of

the second molar of *Ouranopithecus macedoniensis*, it is assumed that its diet consisted of harder foods such as nuts or tubers.

Source (edited): "http://en.wikipedia.org/wiki/Ouranopithecus"

Oxydactylus

Oxydactylus, is an extinct terrestrial herbivorous genus of the tribe *Camelini*, family *Camelidae*, endemic to North America Oligocene through the Middle Miocene (30.8–13.6 mya) and in existence for approximately 17.2 million years.

Oxydactylus is from the Ancient Greek οξύς (oxys, "sharp")and δάκτυλος (daktylos, "finger")

Taxonomy

Oxydactylus was named by Peterson in 1904. Its type is *Oxydactylus longipes*. It was assigned to *Camelidae* by Peterson (1904) and Carroll (1988).

Morphology

Oxydactylus longipes

They had very long legs and necks, and were probably adapted to eating high vegetation, much like modern giraffes. Unlike modern camelids, they had hooves, rather than tough sole-pads and splayed toes.

Four specimens were examined by M. Mendoza, C. M. Janis, and P. Palmqvist for estimated body mass.

- Specimen 1: 183.3 kg (404.1 lb)
- Specimen 2: 115.9 kg (255.5 lb)
- Specimen 3: 116.5 kg (256.8 lb)
- Specimen 4: 126.3 kg (278.4 lb)

Species

O. benedentatus (syn. Procamelus conscerptus), O. cameloides, O. campestris, O. lacota, O. longipes (type species), O. longirostris, O. lulli, O. wyomingensis

Sister genera

- *Aepycamelus* (syn. *Alticamelus, Homocamelus*)
- *Aguascalientia*
- *Alforjas*
- *Australocamelus*
- *Blancocamelus*
- *Camelops*
- *Cuyamacamelus*
- *Eulamaops*
- *Floridatragulus* (syn. *Hypermekops*)
- *Gentilicamelus*
- *Gigantocamelus*
- *Hesperocamelus*
- *Megacamelus*
- *Megatylopus*
- *Michenia*
- *Miotylopus* (syn. *Dyseotylopus*)
- *Nothokemas*
- *Palauchenia*
- *Paracamelus*
- *Paralabis*
- *Paratylopus*
- *Pliauchenia*
- *Poebrodon*
- *Poebrotherium*
- *Priscocamelus*
- *Procamelus*
- *Protolabis*
- *Pseudolabis*
- *Tanymykter*
- *Titanotylopus*

Source (edited): "http://en.wikipedia.org/wiki/Oxydactylus"

Pachyrukhos

Pachyrukhos is an extinct genus of mammal from the Oligocene and Miocene of South America.

It was about 30 cm (1 ft) long and closely resembled a rabbit, possessing a short tail and long hind feet. *Pachyrukhos* was probably also able to hop, and it had a rabbit-like skull with teeth adapted for eating nuts and tough plants. The complexity of its hearing apparatus in the skull suggests that its hearing would have been very good, and that it probably had large ears. It also had large eyes, suggesting that it may have been nocturnal. These similarities are the result of convergent evolution, since, while quite unrelated to modern rabbits, *Pachyrukhos* filled the same ecological niche.

Source (edited): "http://en.wikipedia.org/wiki/Pachyrukhos"

Palaeomeryx

Palaeomeryx is an extinct genus of Artiodactyla, of the family Palaeomerycidae, endemic to Europe and Asia from the Miocene epoch, 16.9—7.25 Ma, existing for approximately 9.65 million years.

Taxonomy

Palaeomeryx was named by von Meyer (1834). It is the type genus of Palaeomerycidae, Palaeomerycinae. It was assigned to Palaeomerycidae by Carroll (1988) and Sach and Heizmann (2001); and to Palaeomerycinae by Prothero and Liter (2007).

Fossil distribution

- Level C1, Quarry Quebra Bilhas, Lisbon, Portugal
- Buñol, Valencia, Spain
- Lufeng, Yunnan Province, China
- Krivoy Rog, Ukraine

Source (edited): "http://en.wikipedia.org/wiki/Palaeomeryx"

Palaeotragus

Palaeotragus ("Ancient Antelope") was a genus of very large, primitive okapi from the Miocene of Africa, Asia, and Europe.

Palaeotragus primaevus is the older species, being found in early to mid-Miocene strata, while *Palaeotragus germaini* is found in Late Miocene strata.

P. primaevus is distinguished from *P. germaini* by having no pair of ossicones. It was also the smaller species, being a little under 2 meters at the shoulders. *P. germaini* had a pair of ossicones, and in life, it would have resembled either a short-necked, 3 meter tall giraffe, or a gargantuan okapi.

Source (edited): "http://en.wikipedia.org/wiki/Palaeotragus"

Paleogale

Paleogale is an extinct genus of cat-like mammal belonging to the suborder Feliformia endemic to North America and Europe exiting from ~35.9—35.8 to 20.4 million years ago (Appearance Event Ordination). Its closest living relatives are the genets, civets, and linsangs.

Taxonomy

Paleogale was assigned to the order Carnivora by von Meyer in 1846 and to the superfamily Viverravoidea by Flynn and Galiano in 1982 and again to the order Carnivora by Baskin in 1998.

Species

P. dorothiae (N. America), P. minuta (N. America & Europe) *P. sanguinarius* (N. America), *P. sectoria* (N. America). *P. sectoria* is synonymous with *Bunaelurus infelix, Bunaelurus lagophagus, and Canis osorum.*

Morphology

According to Legendre and Roth, the body mass of *Paleogale* is 0.90 grams (0.8 pounds) to 0.32 grams (0.5 pounds).

Source (edited): "http://en.wikipedia.org/wiki/Paleogale"

Paleoparadoxia

Paleoparadoxia is a genus of large, herbivorous marine mammals that inhabited the northern Pacific coastal region during the Miocene epoch (20 to 10 million years ago). It ranged from the waters of Japan (Tsuyama and Yanagawa), to Alaska to the north, and down to Baja California, Mexico. It is believed there are only two complete fossil skeletons in the world, one at Fukushima Museum in Fukushima Prefecture Japan and one at the SLAC National Accelerator Laboratory in Northern California.

Paleoparadoxia is thought to have fed primarily on seaweeds and sea grasses. The jaws and the angle of the teeth resemble a backhoe bucket. Its bulky body was well adapted for swimming and underwater foraging, but not for extended deep-sea living or deep diving. Like the modern-day seal, *Paleoparadoxia* probably came on shore for breeding and basking in the sun. It also resembled the Desmostylus.

Gallery

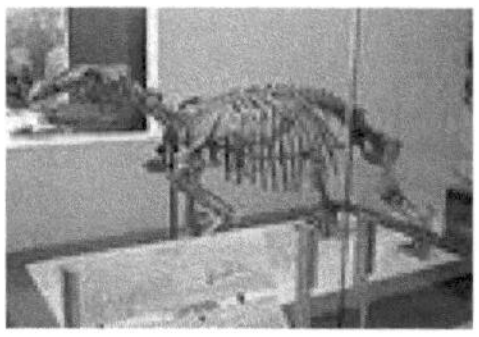

Palaeoparadoxia tabatai

Paleoparadoxia on display at the Natural History Museum of Los Angeles county

Restoration of *Palaeoparadoxia tabatai*

Source (edited): "http://en.wikipedia.org/wiki/Paleoparadoxia"

Palorchestes

Palorchestes ('ancient leaper or dancer') is an extinct genus of terrestrial herbivorous marsupial of the family Palorchestidae. The genus was endemic to Australia, living from the Late Miocene subepoch through the Pleistocene epoch (around 11.6 mya – 11,000 years ago), and thought to be in existence for approximately 11.59 million years.

Description

One species, *Palorchestes azael*, was almost as large as a horse, being around 2.5 metres (8.2 ft) in length, and had four powerful legs. The appearance of the animal's nasal bones suggests that it possessed a short proboscis, leading to the nickname of the "marsupial tapir". Since it is unrelated to tapirs, this similarity in nose shape is an example of convergent evolution. *Palorchestes* front legs bore large claws, similar to those of a koala, which it probably used to pull down leaves and strip the bark from trees.

The long symphysis at the lower jaw of all *Palorchestes* species indicates that the tongue was long and protrudible, like that of a giraffe.

Discovery

Fossilized remains of *Palorchestes azael* have been found at the Naracoorte Caves fossil site in Australia.

Etymology

The generic name was coined by Sir Richard Owen, who first found what he thought was the fragmentary jaw of a prehistoric kangaroo. It was not until more postcranial elements were found did anyone realize that *Palorchestes* was actually a diprodontid, and not a kangaroo.

Sister genera

Ngapakaldia, *Pitikantia*

Source (edited): "http://en.wikipedia.org/wiki/Palorchestes"

Palorchestidae

The family **Palorchestidae** contains four genera with eight species described. All species are extinct.

- *Propalorchestes* (Murray, 1986)
  - *P. novaculocephalus* (Murray, 1990)
  - *P. painei* (Murray, 1986)
- *Ngapakaldia* (Stirton, 1967)
  - *N. bonythoni* (Stirton, 1967)
  - *N. tedfordi* (Stirton, 1967)
- *Palorchestes* (Owen, 1873)
  - *P. selestiae* (Mackness, 1995)
  - *P. paine* (Woodburne, 1967)
  - *P. azeal* (Owen, 1873)
- *Pitikantia dailyi* (Stirton, 1967)

Source (edited): "http://en.wikipedia.org/wiki/Palorchestidae"

Pampatheriidae

Pampatheridae ("Pampas beasts") is an ancient family, now extinct, of large armadillo-like plantigrade armored xenarthrans. They are related to Glyptodontidae, an extinct family of much larger and more heavily armored xenarthrans, as well as to extant armadillos (Dasypodidae). Pampatheres evolved in South America during its long period of Cenozoic isolation. *Holmesina* spread to North America after the formation of the Isthmus of Panama as part of the Great American Interchange. They finally disappeared on both continents in the end-Pleistocene extinctions.

Source (edited): "http://en.wikipedia.org/wiki/Pampatheriidae"

Paraceratherium

Paraceratherium, also commonly known as ***Indricotherium*** or ***Baluchitherium*** (see taxonomic discussion below), is an extinct genus of gigantic hornless rhinoceros-like mammals of the family Hyracodontidae, endemic to Eurasia and Asia during the Eocene to Oligocene 37.2—23.030 Mya, existing for approximately 14.17 million years.. It was first discovered in the Baluchistan province of Pakistan, hence the name, by Sir Clive Forster Cooper.

Background

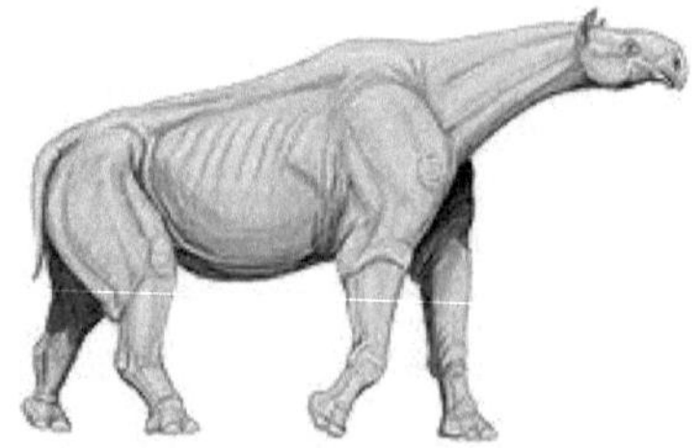

Restoration

Paraceratherium is the largest land mammal known, larger than the largest species of mammoths (*Mammuthus sungari*, which may have approached it in size and weight). It is also known as the "giraffe rhinoceros". Adult *Paraceratherium* are estimated to have been 5.5 metres (18 ft) tall at the shoulder, 9 metres (30 ft) in length from nose to rump, a maximum raised head height of about 8 metres (26 ft), and a skull length of 1.5 metres (4.9 ft). Weight estimates vary greatly, but most realistic and reliable weight estimates are about 20 tonnes. This puts it in the weight range of some medium-sized sauropod dinosaurs.

It was an herbivore that stripped leaves from trees with its down-pointing, tusk-like upper teeth that occluded forward-pointing lower teeth. It had a long, low, hornless skull and vaulted frontal and nasal bones. Its front teeth were reduced to a single pair of incisors in either jaw, but they were conical and so large that they looked like small tusks. The upper incisors pointed straight downwards, while the lower ones jutted outwards. The upper lip was evidently extremely mobile. The neck was very long, the trunk robust, and the limbs long and thick, column-like.

Its type of dentition, its mobile upper lip and its long legs and neck indicate that it was a browser that lived on the leaves and twigs of trees and large shrubs.

Taxonomy

While more distinct at the species level, there is uncertainty and disagreement with regard to the genus level of taxonomy.

Paraceratherium was first described by Forster Cooper in 1911. The genus *Baluchitherium* was first described by Forster Cooper in 1913. The genus *Indricotherium* was first described by Borissiak in 1915.

Baluchitherium is now widely regarded as a synonym of (i.e. the same as) either *Paraceratherium* or *Indricotherium.*

P. bugtiense compared to a man

However, there has been disagreement over whether *Indricotherium* is a distinct genus from *Paraceratherium*. Lucas and Sobus in their 1989 review of the subfamily Indricotheriinae (see reference below), argue for synonymy, and consider that the differences between the two are of species level at most, and may even be the result of sexual dimorphism in a single species, with the larger more robust Indricotherium with larger incisors being probably the male, and the more gracile Paraceratherium the female. Others, however, have expressed doubts about this (concerning the interpretation of the shape of the skull). Even if these two do turn out to be distinct genera, they would still be similar in size and appearance.

If they are considered the same genus, then *Indricotherium* would become a junior synonym of *Paraceratherium*, because, according to the priority principle of scientific classification, the first publication, and hence the oldest valid name, takes priority and the name *Paraceratherium* predates the other.

Here Lucas and Sobus are followed. They consider *Indricotherium*, *Baluchitherium*, *Thaumastotherium* Forster Cooper, 1913a, *Aralotherium* Borissiak, 1939, and *Dzungariotherium* Xu and Wang, 1973 all as junior synonyms of *Paraceratherium.*

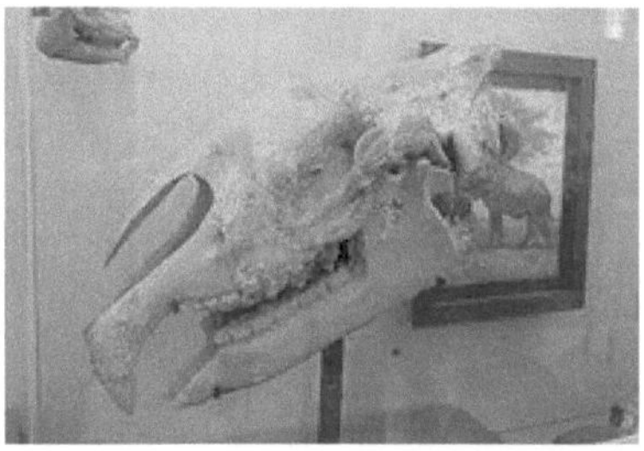

Paraceratherium transouralicum skull

Lucas and Sobus recognise four valid species of *Paraceratherium*. One morc (P. zhajremensis) has been tentatively added. The *Paraceratherium* species are:

Skull of *Paraceratherium bugtiense.*

***Paraceratherium bugtiense* (Pilgrim, 1908)** from the Oligocene of Pakistan is the type species of *Paraceratherium*. *Baluchitherium osborni* Forster Cooper, 1913a is a junior synonym. It was first found in the Chitarwata Formation of the Bugti Hills, Balochistan, after which it was originally named. New specimens of *P. bugtiense* were unearthed in the last decade by a French-Pakistani team (Antoine et al., 2004; Métais et al., 2009).

***Paraceratherium transouralicum* (Pavlova, 1922)**. Also known as *Indricotherium transouralicum*, this is the best known and most widespread species, known from the middle and late Oligocene of Kazakhstan, Mongolia, and Nei Monggol in northern China. Lucas and Sobus list the following species as synonyms: *Baluchitherium grangeri* Osborn, 1923, *Indricotherium asiaticum* Borissiak, 1923, *Indricotherium minus* Borissiak, 1923.

***Paraceratherium orgosensis* (Chiu, 1973)** is the largest species, the teeth being at least a quarter again as big as *P. transouralicum* (see Lucas and Sobus p. 363/fig.19.2). It is known from the middle and late Oligocene of Xinjiang, northwest China. The three synonyms are *Dzungariotherium orgosensis* Chiu, 1973 and (each of the following named after a separate skull) *Dzungariotherium turfanensis* Xu & Wang, 1978 and *Paraceratherium lipidus* Xu & Wang, 1978. While there is some variation in details of the proportions of the skull (perhaps due to sexual dimorphism), all occur in a close geographical region and have distinct first and second upper molar crochets.

***Paraceratherium prohorovi* (Borissiak, 1939)** from the late Oligocene or early Miocene of eastern Kazakhstan.

***Paraceratherium zhajremensis* (Osborn, 1923)** from the Middle and late Oligocene of India.

Other remains referable to *Paraceratherium* or related taxa were found in Southeastern Europe and Asia Minor, notably in Turkey (Antoine et al., 2008).

Paraceratherium means "near horn animal" in old Greek, indicating the fact that is was hornless, but related to the rhinoceroses.

Indricotherium is named after a mythical Russian beast called the Indrik-Beast, considered the most powerful creature and the father of the animals.

Baluchitherium means "beast of Baluchistan", as it was first discovered in Baluchistan, western Province of Pakistan.

In popular culture

Paraceratherium was main focus in the third episode of the popular BBC documentary *Walking with Beasts*. In the programme the narrator always calls the creature an Indricothere, not a specific genus. Although in the accompanying book's information box about the animal, it states that the Indricothere is called *Paraceratherium*.

Source (edited): "http://en.wikipedia.org/wiki/Paraceratherium"

Paracynarctus

Paracynarctus is an extinct genus of the *Borophaginae* and a terrestrial canine (Bear-dog) which inhabited most of North America during the Early Hemingfordian stage of the Early Miocene subepoch through the Early Barstovian stage of the Middle Miocene subepoch 20.6—13.6 Ma, existing for approximately 7 million years.

As with other *Borophaginae, Paracynarctus* was a "bone-crushing dog" with powerful teeth and jaws, and hyena-like features.

Morphology

Fossil specimens of two individuals' body mass were examined by Legendre and Roth. The first specimen was estimated to weigh 8.57 kg (19 lb). The second specimen was estimated to weigh 8.18 kg (18 lb).

Species

- ***P. kelloggi*** (synonymous with *Cynodesmus casei*) was recombined as *Tomarctus kelloggi* several times and again as *Cynarctus kelloggi* by Wang (1994) and again by Wang as *Paracynarctus kelloggi*.
- ***P. sinclairi*** was named by X. Wang.

Fossil distribution

P. kelloggi was originally found Virgin Valley, Nevada in a Barstovian terrestrial horizon.

Source (edited): "http://en.wikipedia.org/wiki/Paracynarctus"

Paradaphoenus

Paradaphoenus is a physically small member of the extinct family of terrestrial carnivores belonging to the suborder Caniformia, family Amphicyonidae ("bear dog"), and which inhabited North America from the Early Oligocene subepoch to the Early Miocene subepoch living 33.9—16.3 Ma, existing for approximately 17.6 million years.

Taxonomy

Paradaphoenus was named by Wortman and Matthew (1899) and was assigned to Daphoeninae by Hunt (1998); and to Amphicyonidae by Wortman and Matthew (1899), Carroll (1988) and Hunt (2001).

Morphology

A single specimen was measured by Legendre and Roth in 1988 and estimated mass was 1.89 kg (4.1 lbs).

Fossil distribution

Haystack, Oregon, Banner County, Nebraska, Dawes County, Nebraska, Sheep Mtn, South Dakota.

Species

P. cuspigerus (synonymous with *Amphicyon entoptychi*), *P. minimus*, *P. tooheyi*

Sister genera

Afrocyon, Amphicyanis, Arctamphicyon, Brachycyon, Cynodictis, Drassonax, Goupilictis, Guangxicyon, Haplocyon, Haplocyonoides, Haplocyonopsis, Harpagophagus, Ictiocyon, Pachycynodon, Protemnocyon, Pseudamphicyon, Pseudarctos, Pseudocyonopsis, Sarcocyon, Symplectocyon, Vishnucyon

Source (edited): "http://en.wikipedia.org/wiki/Paradaphoenus"

Parahippus

Parahippus is an extinct relative of the modern horse, very similar to *Miohippus*, but slightly larger, at around 1 metre (10 h) tall, at the withers.

About 20 million years ago, during the Miocene Period, the Earth was drastically changing: great plains were developing, forests were thinning out, mountains were forming, and swamps were drying up. It is believed these developments affected many changes in the existing horse ancestors.

Description

Parahippus was larger than *Miohippus*, with longer legs and face. The bones in the legs were fused and this, along with muscle development, allowed *Parahippus* to move with forward-and-back strides. Flexible leg rotation was eliminated, so that the animal was better adapted to fast forward running on open ground without moving from side to side. Most importantly, *Parahippus* was able to stand on its middle toe, instead of walking on pads, which gave it the ability to run faster; its weight was supported by ligaments under the fetlock to the big central toe. The side toes were almost vestigial, and seldom touched the ground.

Since leafy food had become scarce, these animals were forced to subsist on the newly-evolved grasses that were by now taking over the plains, and their teeth adapted accordingly. The extra molar crest that was variable in *Miohippus* became permanent in *Parahippus*. The molars developed high crowns and a hard covering for grinding the grass, which was typically covered with high-silica dust and sand.

Source (edited): "http://en.wikipedia.org/wiki/Parahippus"

Parahippus leonensis

Parahippus leonensis is an extinct proto-horse of the family Equidae that was endemic to North America during the Miocene from 23.030—16.3 Ma living for approximately 6.73 million years.

Parahippus leonensis was named for Leon or more specifically Leon County, Florida.

Taxonomy

Parahippus leonensis was named by Sellards (1916). Its type specimen is FGS 5084. Its type locality is Griscom Plantation site, which is in a Miocene marine limestone in the Torreya Formation of Florida. It was recombined as Hippodon leonensis by Quinn (1955); it was considered a nomen dubium by Macdonald (1992).

Morphology

Body mass

Two specimens were examined by M. Mendoza, C. M. Janis, and P. Palmqvist for body mass. The results were:

- Specimen 1: 123.7 kg (270 lb)
- Specimen 2: 48.7 kg (110 lb)

Origin

Parahippus leonensis was the next step in evolution after *Miohippus*. *Parahippus* means "side horse" and has been called the evolutionary link between the older forest-dwelling horses and modern plains-dwelling grazers. It is believed to be a close relative to the group from which modern horses evolved. Side may refer to side branches on the posterior crest of the upper molars which separated *Parahippus* from *Anchitherium*.

This genus of horses had a long head with eyes situated back from the middle of the skull. It had three toes, like other primitive horses, however *Parahippus leonensis* had smaller side toes. It was a common species from the Great Plains to Florida. *Parahippus leonensis* weighed in at about 72.5 Kg (160 pounds).

Parahippus leonensis was very likely the prey of *Amphicyon* or Bear-dog, and dog-like *Temnocyon*.

Source (edited): "http://en.wikipedia.org/wiki/Parahippus_leonensis"

Paramachairodus

Paramachairodus is an extinct genus of saber-tooth cat of the subfamily Machairodontinae endemic to Europe and Asia, during the late Miocene from 15 to 9 Ma.

Paramachairodus is one of the oldest known true saber-toothed cats. A large number of fossils were discovered in Cerro Batallones, a Late Miocene fossil site near Madrid, Spain. Two leopard-sized species are known, *Paramachairodus ogygia* from Vallesian to early Turolian age and *Paramachairodus orientalis* from the Turolian. A third species *Paramachairodus maximiliani*, is considered a synonym of *Paramachairodus orientalis* by most authors .

The animal was about 58 cm high at the shoulder, similar to a leopard, but with a more supple body. The shape of its limbs suggests that it may have been an agile climber, and could have hunted relatively large prey.

Source (edited): "http://en.wikipedia.org/wiki/Paramachairodus"

Parapropalaehoplophorus

Parapropalaehoplophorus septentrionalis was a relatively small (compared to, say, *Glyptodon*) species of glyptodont, extinct relatives of the modern armadillo. The mammal, identified in 2007 from the fossilized remains of a specimen found in 2004, weighed approximately 200 pounds and had a shell covered by tiny circular bumps. It lumbered around northern Chile in an area now dominated by the Andes mountain range, some 18 million years ago.

Source (edited): "http://en.wikipedia. org/wiki/Parapropalaehoplophorus"

Paratetralophodon

Paratetralophodon is an extinct genus of gomphothere. *Paratetralophodon* may be an ancestor of *Tetralophodon.* The elephant-like animal lived through the Miocene and Pliocene epochs. It had four tusks and a trunk.

Source (edited): "http://en.wikipedia. org/wiki/Paratetralophodon"

Paratoceras

Paratoceras is a medium-sized extinct genus of Artiodactyla, of the family Protoceratidae, endemic to North America from the Miocene epoch, 16.3—15.97 Ma, existing for approximately 0.33 million years.

Taxonomy

Paratoceras was named by Frick (1937). Its type is Paratoceras macadamsi. It was assigned to Protoceratidae by Frick (1937) and Carroll (1988); and to Protoceratinae by Webb (1981), Prothero (1998), Webb et al. (2003) and Prothero and Ludtke (2007).

Morphology

Poabromylus resembled deer. However they were more closely related to camelids. In addition to having horns in the more usual place, protoceratids had additional, rostral horns above the orbital cavity. *Poabromylus* was larger than Eocene members of Tylopoda: Heteromeryx, Leptoreodon, Leptotragulus, Toromeryx, Trigenicus, and Poabromylus.

Body mass

A single fossil specimen of *Poabromylus* was measured by M. Mendoza, C. M. Janis, and P. Palmqvist for body mass. The specimen was determined to weigh:

- Specimen 1: 73.3 kg (160 lb)

Fossil distribution

Fossils have been recovered from:

- Gaillard Cut, Panama
- Balumtum Sandstone Formation, Chiapas, Mexico
- Suchilquitongo Formation, Oaxaca, Mexico
- Trinity River Pit 1, Fleming Formation, San Jacinto County, Texas

Source (edited): "http://en.wikipedia. org/wiki/Paratoceras"

Paratomarctus

Paratomarctus an extinct member of the *Borophaginae*, subtribe *Borophagina*, a terrestrial canine (Bear-dog) which inhabited most of North America from the Harrisonian stage to Early Barstovian stage of the Miocene epoch living 16.3—5.3 mya, existed for approximately 11 million years.

Canid competitors

Paratomarctus was one of the last of the Borophaginae and shared its North American habitat with other canidae: Borophagus (23.3—3.6 Mya), Epicyon (20.6—5.330 Ma), Carpocyon (20.4—3.9 Ma), Aelurodon (23.03—4.9 Ma), and the first emerging wolf, Canis lepophagus appearing 10.3 Ma.

Taxonomy

Paratomarctus was named and assigned to *Borophagina* by Wang et al. (1999). It was recombined as *Tephrocyon* temerarius by Matthew and Cook (1909), Peterson (1910), Merriam (1913), Matthew (1918) and Merriam (1919); it was recombined as Tomarctus temerarius by Matthew (1924), Green (1948), Galbreath (1953), Downs (1956), Sutton (1977), Voorhies (1990) and Munthe (1998); it was recombined as Paratomarctus temerarius by Wang et al. (1999).

Morphology

Two specimens were examined by Legendre and Roth for body mass. The first specimen was estimated to weigh 12.9 kg (28 lb). The second specimen was estimated to weigh 12 kg (26 lb).

Fossil range

Fossils were uncovered throughout most of the western United States.

Source (edited): "http://en.wikipedia. org/wiki/Paratomarctus"

Paroreodon

Paroreodon is an extinct genus of oreodont of the family *Merycoidodontidae*, subfamily *Merycoidodontinae*, endemic to North America during the Oligocene-Miocene subepochs (30.8—20.6 mya), existing for approximately 10.2 million years.

Taxonomy

Paroreodon was named by Thorpe (1921). Its type is *Paroreodon marshi.* It was synonymized subjectively with *Oreodontoides* by Schultz and Falkenbach in 1947 and assigned to *Merycoidodontidae* by Thorpe (1921) and Lander (1998).

Fossil distribution

Wheeler County, Oregon.

Species

Paroredon parvus (synonymized with *Oreodontoides stocki, Paroredon marshi*)

Source (edited): "http://en.wikipedia.org/wiki/Paroreodon"

Pediomeryx

Pediomeryx is an extinct genus of Artiodactyla, of the family Palaeomerycidae, endemic to North America from the early Miocene epoch (Hemingfordian stage) 10.3—4.9 Ma, existing for approximately 5.4 million years.

Taxonomy

Pediomeryx was named by Stirton (1936). It was assigned to Dromomerycidae by Stirton (1936) and Janis and Manning (1998); to Dromomerycinae by Hulbert and Whitmore (2006); and to Cranioceratini by Prothero and Liter (2007).

Morphology

Body mass

A single fossil specimen of *Pediomeryx* was measured by M. Mendoza, C. M. Janis, and P. Palmqvist for body mass.

- Specimen 1: 251 kg (550 lb)

Fossil distribution

Fossils have been recovered from the Midway Site in Florida, Saskatchewan, Boron, California, and several sites in Nebraska and Wyoming.
Source (edited): "http://en.wikipedia.org/wiki/Pediomeryx"

Peltephilus

Peltephilus ferox, the **Horned Armadillo**, was a species of dog-sized, armadillo-like xenarthran mammal which first inhabited Argentina during the Oligocene epoch, and became extinct in the Miocene epoch. Notably, the scutes on its head were so developed that they formed horns protecting its eyes. Aside from the horned gophers of North America, it is the only known fossorial horned mammal.

Although it had traditionally been perceived as a carnivore because of its large, triangular-shaped teeth, Vizcaino and Farina argued in 1997 that *Peltephilus* was a herbivore.
Source (edited): "http://en.wikipedia.org/wiki/Peltephilus"

Peraceras

Peraceras is an extinct genus of rhinoceros endemic to North America during the Miocene, living from 20.6—10.3 mya and existing for approximately 10. 3 million years.

Taxonomy

Peraceras was named by Cope (1880). Its type is *Peraceras superciliosum*. It was assigned to Rhinocerotidae by Cope (1880) and Carroll (1988); and to Aceratheriinae by Prothero (1998).

Morphology

Four specimens were examined by M. Mendoza, C. M. Janis, and P. Palmqvist as well as M. Fortelius and J. Kappelman for estimated body mass with the results as follows:

- Specimen 1: 98.7 kg (220 lb)
- Specimen 2: 315.7 kg (700 lb)
- Specimen 3: 408.6 kg (900 lb)
- Specimen 4: 121.7 kg (270 lb)

Source (edited): "http://en.wikipedia.org/wiki/Peraceras"

Percrocuta

Percrocuta is an extinct genus of hyena-like feliform carnivores. It lived in Europe, Asia, and Africa, during the Miocene epoch.

With a maximum length of 1.50 m (5 ft), *Percrocuta* was much bigger than its modern relatives, but smaller than a female lion. Like the Spotted Hyena, *Percrocuta* had a robust skull and powerful jaws. Similar to modern hyenids, its hind legs were shorter than the front legs, resulting in a characteristic sloping back.

Percrocuta was introduced as a genus of Percrocutidae in 1938. Percrocuta's relation to the Hyaenidae family was debated until 1985, when Percrocuta, Dinocrocuta, Belbus, and Allohyaena were accepted as the four genera of Percrocutidae.

Fossil evidence

P. abessalomi is known only from a skull, two mandibles, and two teeth. These fossils were all collected the Belomechetskaja, Georgia area and date from the sixth mammal neogene (MN) zone. This species is the best known of the Percrocutidae family. P. miocenica is known from only a few mandibles, found in Yugoslavia and Turkey. These fossils also date from 6 MN.

Source (edited): "http://en.wikipedia. org/wiki/Percrocuta"

Percrocutidae

The **Percrocutidae** form an extinct family of hyena-like feliform carnivores endemic to Asia, Africa, and Southern Europe from the Miocene through Pliocene living 20—2.59 Ma existing for approximately 17.41 million years.

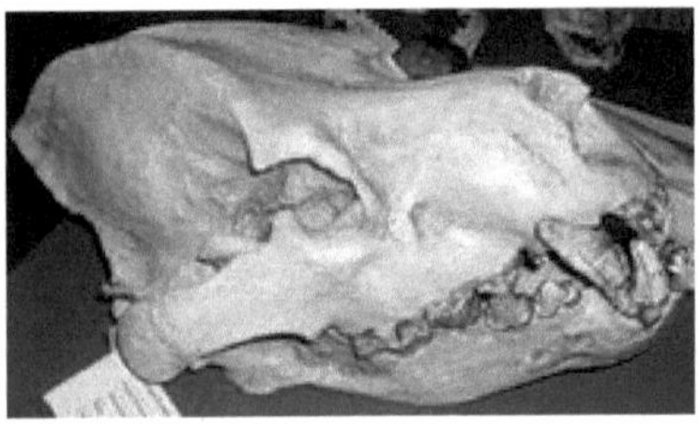

Dinocrocuta gigantea skull cast, Zoologisk Museum.

The first percrocutids are known from the middle Miocene of Europe and western Asia and belonged to the genus *Percrocuta*. *Percrocuta* already had large premolars, but did not carry such a massive bite as the later form *Dinocrocuta*, from the later Miocene. Originally, these carnivores were placed with the hyenas in the family Hyaenidae. Today, most scientists consider Percrocutidae to be a distinct family—although sometimes it is placed with carnivoran genera such as *Stenoplesictis* into the family Stenoplesictidae.

Genera of Percrocutids

- *Percrocuta* (including Capsatherium; Middle Miocene to Late Pliocene of Africa, Middle to Late Miocene of Eurasia)
- *Dinocrocuta* (Middle Miocene of Africa, Middle to Late Miocene of Asia)

The list follows McKenna and Bells *Classification of Mammals* for prehistoric genera (1997). In contrast to McKenna and Bell's classification they are not included as a subfamily into the Hyaenidae but as a separate family Percrocutidae.

Source (edited): "http://en.wikipedia. org/wiki/Percrocutidae"

Phenacocoelus

Phenacocoelus is an extinct genus of oreodont of the family *Merycoidodontidae*, subfamily *Merycoidodontinae*, endemic to North America during the Late Oligocene-Early Miocene subepochs (24.8—20.6 mya), existing for approximately 4.2 million years.

Taxonomy

Phenacocoelus was named by Peterson (1907). Its type is *Phenacocoelus typus*. It was assigned to *Merycoidodontidae* by Peterson (1907) and Lander (1998).

Morphology

A single specimen was examined by M. Mendoza for body mass and was estimated to have a weight of 73.0 kg (160.9 lbs).

Fossil distribution

Fossils have been uncovered in eastern Wyoming and western Nebraska

Species

Phenacocoelus typus (syn. *P. munroensis*).

Source (edited): "http://en.wikipedia. org/wiki/Phenacocoelus"

Phlaocyon

Phlaocyon ("crushing dog") is an extinct genus of the *Borophaginae* and a terrestrial canine which inhabited most of North America during the Whitneyan stage of the Early Oligocene through Late Hemingfordian stage of the Early Miocene epoch 33.3—16.3 Mya existing for approximately 17.3 million years.

Taxonomy

Phlaocyon was about 80 centimetres (31 in) in body length, and looked more like a cat or raccoon than a dog, but its skull anatomy shows it to be a primitive canine. *Phlaocyon* probably lived like a raccoon, often climbing trees. Its head was short, wide, and had forward-facing eyes. Unlike modern canines, *Phlaocyon* had no specialised teeth for slicing flesh. It is thought to have been an omnivore.

Morphology

Body mass

Two specimens were examined by Legendre and Roth for body mass.

- Specimen 1 was estimated to weigh 1.81 kg (4.0 lb).
- Specimen 2 was estimated to weigh 1.87 kg (4.1 lb).

Fossil distribution

- Foree Site, John Day Formation, Wheeler County, Oregon (P. latidens) ~30.8—20.6 Ma.
- Brooksville 2 Site, Hernando County, Florida (P. taylori) ~30.8—20.6 Ma.
- SB-1A Live Oak Site, Suwannee County, Florida (P. leucosteus) ~24.8—20.6 Ma.
- Buda Mine, Alachua County, Florida (P. indent) ~24.8—20.6 Ma.
- Wewela Site, Tripp County, South Dakota (P. minor) ~26.3—24.8 Ma.

Species

- *Phlaocyon achoros*
- *Phlaocyon annectens*
- *Phlaocyon latidens*
- ***P. leucosteus***
- *Phlaocyon mariae*
- *Phlaocyon marslandensis*
- *Phlaocyon minor*
- *Phlaocyon multicuspus*
- *Phlaocyon taylori*
- *Phlaocyon yatkolai*

Source (edited): "http://en.wikipedia.org/wiki/Phlaocyon"

Phlaocyon achoros

Phlaocyon achoros is an extinct species of the genus Phlaocyon, belonging to the subfamily Borophaginae and tribe Phlaocyonini, a canid which inhabited the southeastern North America from the Late Oligocene to Miocene living 24.6—20.8 mya and existed for approximately 4.2 million years.

Taxonomy

Phlaocyon achoros was named by Frailey (1979). Its type locality is Buda Mine, which is in a Harrisonian sinkhole horizon in Florida. It was recombined as *Phlaocyon achoros* by Wang et al. (1999) and Hayes (2000).

Morphology

Body mass

Two specimens were examined by Legendre and Roth for body mass.

- Specimen 1 was estimated to weigh 1.45 kg (3.2 lb).
- Specimen 2 was estimated to weigh 1.52 kg (3.4 lb).

Fossil distribution

- Buda Mine Site, Alachua County, Florida ~24.8—20.6 Ma.

Source (edited): "http://en.wikipedia.org/wiki/Phlaocyon_achoros"

Phlaocyon annectens

Phlaocyon annectens is an extinct species of the genus Phlaocyon, belonging to the subfamily Borophaginae and tribe Phlaocyonini, a canid endemic to central and western North America from the Late Oligocene to Early Miocene living 24.6—20.8 mya and existed for approximately 4.2 million years.

Taxonomy

Phlaocyon annectens was named by Peterson (1907). Its type locality is Beardog Hill, which is in a Harrisonian fluvial sandstone in the Upper Harrison Beds Formation of Nebraska. It was recombined as *Phlaocyon annectens* by Vanderhill (1980) and Wang et al. (1999).

Morphology

Body mass

Two specimens were examined by Legendre and Roth for body mass.

- Specimen 1 was estimated to weigh 1.81 kg (4.0 lb).
- Specimen 2 was estimated to weigh 1.87 kg (4.1 lb).

Fossil distribution

- Castolon (TMM 40635), Brewster County, Texas ~24.8—20.6 Ma.
- American Museum-Cook Quarry, Sioux County, Nebraska ~24.8—20.6 Ma.
- Van Tassel, Niobrara County, Wyoming ~24.8—20.6 Ma.
- Beardog Hill, Upper Harrison Beds Formation, Sioux County, Nebraska ~24.8—20.6 Ma.

Source (edited): "http://en.wikipedia.org/wiki/Phlaocyon_annectens"

Phlaocyon latidens

Phlaocyon latidens is an extinct species of the genus Phlaocyon, belonging to the subfamily Borophaginae and tribe Phlaocyonini, a canid which inhabited southeastern North America from the Late Oligocene to Miocene living 33.3—20.6 mya and existed for approximately 12.7 million years.

Taxonomy

Phlaocyon latidens was originally named *Galecynus latidens* by Cope (1881). It was recombined as *Cynodictis latidens* by Scott (1898); it was recombined as *Nothocyon latidens* by Matthew (1899), Merriam (1906), Matthew (1907), Peterson (1907), Thorpe (1922), Hall and Martin (1930), Macdonald (1963) and Macdonald (1970); it was recombined as *Cormocyon latidens* by Wang and Fremd (1994); it was recombined as *Phlaocyon latidens* by Xiaoming Wang, Richard H. Tedford, et al. (1999).

Morphology

Body mass

Two specimens were examined by Legendre and Roth for body mass.

- Specimen 1 was estimated to weigh 1.52 kg (3.4 lb).
- Specimen 2 was estimated to weigh 1.58 kg (3.5 lb).

Fossil distribution

- North Blue Basin Site, John Day Formation, Grant County, Oregon

~33.3—30.8 Ma.

- Foree Site, John Day Formation, Wheeler County, Oregon ~30.8—20.6 Ma.

Source (edited): "http://en.wikipedia.org/wiki/Phlaocyon_latidens"

Phlaocyon mariae

Phlaocyon mariae is an extinct species of the genus Phlaocyon, belonging to the subfamily Borophaginae and tribe Phlaocyonini, a canid which inhabited central western North America from the Miocene living 20.6—16.3 Ma., and existed for approximately 4.3 million years.

Taxonomy

Phlaocyon mariae was named by Matthew (1899). It was assigned to Canidae by Matthew (1899) and Carroll (1988); and to Phlaocyonini by Wang et al. (1999).

Fossil distribution

- Aletomeryx Quarry Site, Runningwater Formation, Cherry County, Nebraska ~20.6—16.3 Ma.

Source (edited): "http://en.wikipedia.org/wiki/Phlaocyon_mariae"

Phlaocyonini

Phlaocyonini † is a hypocarnivorous clade or tribe of Borophaginae. This is an extinct group of canids that were endemic to North America and which lived during the Oligocene epoch (Whitneyan stage) to the Miocene (Early Barstovian) living ~33.3–5.3 Ma, existing for approximately 28.3 million years.

Phlaocyonini

The clade includes *Cynarctoides* and *Phlaocyon*. These two genus members represent differing approaches of hypocarnivory. The *Cynarctoides* trend toward having molar teeth with crowns formed of crescent-shaped cusps and remain small in size. *Phlaocyon* specialize toward rounded cusps on the molar teeth but of increasing size, with an unusual trend toward being hypercarnivores by two terminal species in the clade.

Four transitional species of *Cormocyon* and *Desmocyon* occupy intermediate positions between the *Phlaocyonini* and *Cynarctina*. These represent a gradual size increase toward medium-size individuals.

Fossil distribution

A few of many sites:

- Wewela Site, Turtle Butte Formation, Tripp County, South Dakota (Phlaocyon minor) ~30.8—20.6 Ma.
- Brooksville 2 Site, Hernando County, Florida (Phlaocyon taylori) ~26.3—24.8 Ma.
- Split Rock Site, Arikaree Formation, Fremont County, Wyoming, (Cynarctoides) ~20.6—16.3 Ma.
- Cedar Run Site, Oakville Formation, Washington County, Texas (Phlaocyon) ~20.3—5.3 Ma.
- Nambe Site, Tesuque Formation, Santa Fe County, New Mexico, (Cynarctoides acridens) ~20.3—5.3 Ma.

Source (edited): "http://en.wikipedia.org/wiki/Phlaocyonini"

Phoberocyon

Phoberocyon is a large extinct genus of carnivorous Ursidae, subfamily Hemicyonidae, endemic to what would be North America during the Miocene living from 20.6—16.3 mya, existing for approximately 4.3 million years.

Taxonomy

Phoberocyon was named by Ginsburg (1955). It was assigned to Ursidae by Ginsburg (1955) and Carroll (1988); and to Hemicyoninae by Hunt (1998).

Morphology

Two specimens were examined by Legendre and Roth for body mass.

- Specimen 1: 83 kg (180 lb).
- Specimen 2: 689.3 kg (1,500 lb).

Fossil distribution

- Thomas Farm Site, Alachua Formation, Gilchrist County, Florida *Phoberocyon johnhenryi* ~20.6—16.3 Ma. This is the only known specimen.

Species

- *Phoberocyon hispanicus* Ginsburg 7 Morales, 1998
- *Phoberocyon dehmi* (Ginsburg, 1955)
- *Phoberocyon aurelianensis* (Mayet, 1908)
- *Phoberocyon youngi*
- *Phoberocyon johnhenryi* (White, 1947)
- *Phoberocyon huerzeleri* Ginsburg, 1955

Source (edited): "http://en.wikipedia.org/wiki/Phoberocyon"

Phoberogale

Phoberogale is an extinct genus of bear-like carnivores of the Oligocene and Miocene epoch, found in France (Europe), California (North America), and Pakistan (Asia), living from around 25 to 18 million years ago.

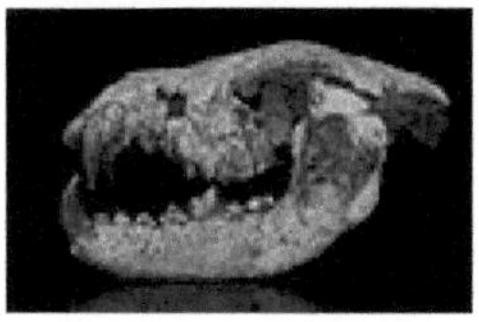

Skull and lower jaw of *Phoberogale shareri*, holotype, OCPC 21794 and 21795.

Taxonomy

Phoberogale was named by Ginsberg and Morales in 1995. Its type species is *Phoberogale bugtiensis* (Cooper, 1923). *Phoberogale shareri* was named after Kevin Sharer, Chairman of the Board of the Natural History Museum of Los Angeles County by Xiaoming Wang, Robert Hunt Jr, Richard Tedford and E. Bruce Lander in 2009. This species belongs to the *Phoberogale–Phoberocyon* lineage and is distinguished from species of *Cephalogale* (s.s.) by its larger size, less-reduced pre-carnassial premolars,and less-elongated M2.

Source (edited): "http://en.wikipedia.org/wiki/Phoberogale"

Phoberomys pattersoni

Phoberomys pattersoni is an extinct rodent that lived in the ancient Orinoco River delta approximately 8 million years ago. It was the second-largest of the roughly 7 species of its genus. Like many other rodents, *Phoberomys* was a herbivore with high-crowned premolars and molars. Informal names include **Ratzilla** and **Mighty Mouse**.

An almost complete skeleton of the giant rodent was discovered in Urumaco, Venezuela in 2000. The new species was later classified with the name *Phoberomys pattersoni* in honor of palaeontologist Brian Patterson. From the fossil, researchers have been able to reconstruct its size and probable lifestyle. It was 3 m (9.8 ft) long, with an additional 1.5 m (4.9 ft) tail, and probably weighed around 700 kg (1,500 lb), making it for some years the largest known rodent for which a good size and weight estimate was possible. Its congener *Phoberomys insolita* was a bit larger still, but it is not known from any reasonably complete remains and thus its size cannot be estimated more precisely. In early 2008, the discovery of *Josephoartigasia monesi* was announced, which was even larger.

Source (edited): "http://en.wikipedia.org/wiki/Phoberomys_pattersoni"

Pierolapithecus

Pierolapithecus catalaunicus is an extinct species of primate which lived about 13 million years ago during the Miocene in what is now Hostalets de Pierola, Catalonia (Spain) giving the name to the species. It is believed by some to be a common ancestor of both modern humans and the other great apes, or at least a species that brings us closer to a common ancestor than any previous fossil discovery.

The species was described by a team of Spanish paleoanthropologists led by Salvador Moyà-Solà on the basis of a fossil specimen discovered in December 2002. The finding was first reported in the journal Science on November 19, 2004.

Pierolapithecus had special adaptations for tree climbing, just as humans and other great apes do: a wide, flat ribcage, a stiff lower spine, flexible wrists, and shoulder blades that lay along its back. Monkeys and lesser apes show more generalized characteristics. The ape does have more primitive, monkeylike features, however, such as a sloped face and short fingers and toes.

The hypothesis that this new species was an ancestor of all modern great apes is controversial due to its location in Spain, since all the existing great ape species live in either Southeast Asia or Africa, and Africa has been the location for so much of the evolution of great apes and humans. However, the Mediterranean Sea expanded and contracted frequently in the past, permitting the dispersal of life between Africa and Europe and the *Pierolapithecus* could have lived on both continents.

Rather than a full common ancestor, it has been suggested that the species may be ancestral to humans, chimpanzees and gorillas but not orangutans, given certain characteristics of the face.

Source (edited): "http://en.wikipedia.org/wiki/Pierolapithecus"

Platybelodon

Platybelodon ("flat-spear tusk") was a genus of large herbivorous mammal related to the elephant (order Proboscidea). It lived during the Miocene Epoch, about 15-4 million years ago, and ranged over Africa, Europe, Asia and North America. Although it thrived during its time, it did not survive past the Miocene.

Palaeobiology

Platybelodon was previously believed to have fed in the swampy areas of grassy savannas, using its teeth to shovel up aquatic and semi-aquatic vegetation. However, wear patterns on the teeth suggest that it used its lower tusks to strip bark from trees, and may have used the sharp incisors that formed the edge of the "shovel" more like a modern-day scythe, grasping branches with its trunk and rubbing them against the lower teeth to cut it from a tree.

Skull of Platybelodon

Platybelodon was very similar to *Amebelodon*, another, closely related gomphothere genus. Due to the shape of the two lower teeth, which are worn by many gomphothere genera (such as *Platybelodon*, *Archaeobelodon*, and *Amebelodon*), they are popularly known as "shovel tuskers."

Platybelodon Molar

Source (edited): "http://en.wikipedia.org/wiki/Platybelodon"

Plesictis

Plesictis is an extinct prehistoric genus of mustelid (originally described as a procyonid) endemic to Europe during the Oligocene 33.9—28.4 Ma existing for approximately 5.5 million years.

Plesictis was a 75 centimetres (2.46 ft) long animal, resembling a weasel with large eyes, or possibly a cacomistle. Its large eyes and very long tail suggest that it may have been nocturnal and arboreal. Judging from its teeth, it was an omnivore.
Source (edited): "http://en.wikipedia.org/wiki/Plesictis"

Plesippus

Plesippus is often considered an intermediary stage between *Dinohippus* and the present day genus, *Equus*.

Mounted skeleton of Hagerman Horse (*Equus simplicidens*)

The famous fossils found near Hagerman, Idaho were originally thought to be a part of the genus *Plesippus* Hagerman Fossil Beds (Idaho), a Pliocene site, dating about 3.5 Ma ago. The fossilized remains were originally called *Plesippus shoshonensis* but further study by paleontologists determined that fossils represented the oldest remains of the genus *Equus*. Their estimated average weight was 425 kg, roughly the size of an Arabian horse.

At the end of the Pliocene, the climate in North America began to cool down significantly and most of the animals were forced to move south. One group of the *Plesippus* species moved across the land bridge around the Bering Strait into Eurasia around 2.5 Ma ago.
Source (edited): "http://en.wikipedia.org/wiki/Plesippus"

Pliohippus

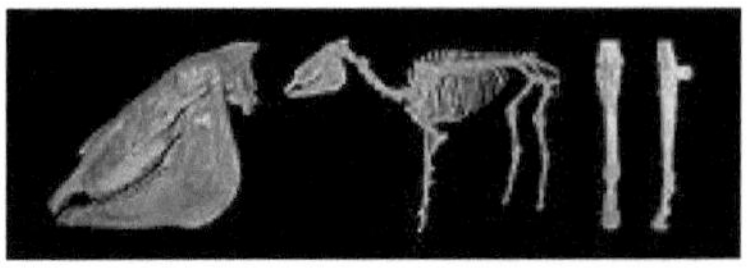

Specimen at American Museum of Natural History

Pliohippus is an extinct genus of

Equidae, the "horse family". *Pliohippus* arose in the middle Miocene, around 12 million years ago, probably from *Calippus*. It was similar in appearance to *Equus*, but had two long extra toes on both sides of the hoof, externally barely visible as callused stubs. The long and slim limbs of *Pliohippus* reveal a quick-footed steppe animal.

Until recently, because of its many anatomical similarities, *Pliohippus* was believed to be the ancestor of the present-day horse and its relatives in *Equus*. Although *Pliohippus* clearly is an equid and thus related to *Equus*, its skull had deep facial *fossae*, a feature not found in any member of *Equus*. Additionally, its teeth were strongly curved, unlike the very straight teeth of modern horses. Consequently, it is unlikely to be the ancestor of the modern horse; instead, it is likely to be the ancestor of *Astrohippus*.

Source (edited): "http://en.wikipedia.org/wiki/Pliohippus"

Plionarctos

Plionarctos is an extinct genus of mammals of the family Ursidae (bears) endemic to North America and Europe during Miocene through Pleistocene, living from ~10.3—3.3 Ma, existing for approximately 7.1 million years.

Indarctos (10.7—9.2 Ma) preceded *Plionarctos* by only a few thousand years and was a contemporary of that bear and shared its habitat. *Plionarctus* preceded and was also contemporary with *Tremarctos floridanus* (4.9 Ma.—11,000 years ago) and shared its habitat.

Plionarctos is the oldest known genus within the subfamily of the short-faced bears (Tremarctinae) endemic to the Americas, and is believed to be ancestral to the clade.

Taxonomy

Plionarctos was named by Frick (1926). Its type is *Plionarctos edensis*. It was assigned to Ursidae by Frick (1926) and Carroll (1988); and to Tremarctini by Hunt (1998).

Morphology

Body mass

Two specimens were examined by Legendre and Roth for body mass.

- Specimen 1 estimated to weigh: 165.5 kg (360 lb)
- Specimen 2 estimated to weigh: 25.3 kg (56 lb)

Fossil distribution

Sites and specimen ages:

- Ile de Ratonneau Breccia, Provence, France ~800,000—100,000 years ago.
- Fort Green Mine, Polk County, Florida paleontological sites ~10.3—4.9 Ma.
- Taunton site, Adams County, Washington *(P. harroldorum)* ~4.9—1.8 Ma. (*Plionarctos harroldorum*)
- Pipe Creek Sinkhole, Grant County, Indiana *(P. edensis)* ~10.3—1.8 Ma.
- Palmetto Mine, Polk County, Florida ~7.9—7.8 Ma.
- Gray Fossil Site, Washington County, Tennessee ~7.0-4.5 Ma.

Source (edited): "http://en.wikipedia.org/wiki/Plionarctos"

Pliopedia

Pliopedia pacifica is an extinct species of walrus found in what is now Central Valley, California, which lived during the late Miocene.

Source (edited): "http://en.wikipedia.org/wiki/Pliopedia"

Pliopithecidae

The family **Pliopithecidae** is the earliest known family of fossil apes. They originated in Africa, and subsequently spread to Europe, before becoming extinct about 10 million years ago. Like modern gibbons, they were adapted to living in the tree tops of dense forests. Their anatomy combined primitive features such as a small braincase, a long snout and a tail in some species. At the same time, they possessed more advanced features such as stereoscopic vision and ape-like teeth and jaws, clearly distinguishing them from monkeys.

Source (edited): "http://en.wikipedia.org/wiki/Pliopithecidae"

Pliopithecus

Pliopithecus antiquus jaw seen from above

Pliopithecus is a genus of extinct primates of the Miocene and Pliocene. It was discovered in 1837 by Édouard Lartet (1801–1871) in France, with fossils subsequently discovered in Switzerland and Spain.

Pliopithecus' had a similar size and form to modern gibbons, to which it may be related, although it is probably not a direct ancestor. It had long limbs, hands, and feet, and may have been able to brachiate, swinging between trees using its arms. Unlike gibbons, it had a short tail, and only partial stereoscopic vision.

Anapithecus is a close relative and was initially considered a subgenus of *Pliopithecus*.

Source (edited): "http://en.wikipedia.org/wiki/Pliopithecus"

Plioviverrops

Plioviverrops is an extinct genus of terrestrial carnivore of the family *Hyaenidae,* endemic to Southern Europe during the Late Miocene subepoch (11.6—5.3 mya) existing for approximately 6.3 million years.

Taxonomy

Plioviverrops was named by Kretzoi (1938). It was assigned to *Hyaenidae* by Flynn (1998).

Sister genera

Adcrocuta (syn. *Hyena*), *Allohyaena, Chasmaporthetes* (syn. *Ailuraena*), *Crocuta, Hyaena, Hyaenictis, Hyaenictitherium, Ictitherium, Leecyaena, Lycyaena, Miohyaena, Pachycrocuta, Palinhyaena, Pliocrocuta, Protictitherium, Thalassictis* (syn. *Palhyaena*), *Tungurictis.*

Source (edited): "http://en.wikipedia.org/wiki/Plioviverrops"

Plithocyon

Plithocyon is an extinct genus of bear-like hemicyonid carnivore of the Miocene epoch, endemic to North America and Europe living from ~15.97—7.25 Ma, existing for approximately 8.72 million years.

Taxonomy

Plithocyon was named by Ginsburg (1955). It was assigned to Ursidae by Ginsburg (1955) and Carroll (1988); and to Hemicyoninae by Hunt (1998).

Morphology

Body mass

Two specimens were examined by Legendre and Roth for body mass.

- Specimen 1: 79.3 kg (170 lb)
- Specimen 2: 652.6 kg (1,400 lb)

Fossil distribution

Sites and age of specimens:

- Hemicyon Quarry, Barstow Formation, San Bernardino County, California ~13.7—11.6 Ma.
- Ruby River Basin No.5, Madison County, Montana ~23—5.3 Ma.
- Pasalar site, Bursa, Turkey ~16—13.7 Ma.
- Pontigne site, France ~11.6—7.2 Ma.

Species

- *Plithocyon antunesi* Ginsburg & Morales, 1998
- *Plithocyon armagnacensis* Ginsburg, 1955
- *Plithocyon barstowensis* (Frick, 1926)
- *Plithocyon bruneti* Ginsburg, 1980
- *Plithocyon conquense* Ginsberg
- *Plithocyon ursinus* (Cope, 1875)

Source (edited): "http://en.wikipedia.org/wiki/Plithocyon"

Pratifelis martini

Pratifelis martini is an extinct feline species that lived in North America during the late Miocene period. A jawbone from the animal was first discovered by H. T. Martin in Wallace County, Kansas, in 1911, and paleontologist

Claude W. Hibbard documented it as a new species.

P. martini, a short-faced cat, was likely bigger than the modern cougar.

Source (edited): "http://en.wikipedia.org/wiki/Pratifelis_martini"

Prepotherium

Prepotherium is an extinct genus of ground sloths that lived during the Miocene period. Fossils of them have been found in Argentina and Venezuela. Examples:

- Prepotherium potens (Ameghino 1891)
- Prepotherium venezuelam (Collins 1934)

Prepotherium differed from Megatherium in being smaller and having a less exaggeratedly convex inferior border of the lower jaw.

Source (edited): "http://en.wikipedia.org/wiki/Prepotherium"

Primelephas

Primelephas, is a genus of Elephantidae that existed during the Miocene and Pliocene epochs. The name of the genus suggests 'first elephant'. These primitive elephantids are hypothesised to be the common ancestor of *Mammuthus*, the mammoths, and the closely allied genera *Elephas* and *Loxodonta*, the African and Eurasian elephants, diverging some four to six million years ago. Unusual for an elephant, it had four tusks. The type species, *Primelephas gomphotheroides*, was described by Vincent Maglio in 1970, the specific epithet indicating the fossil specimens were gomphothere-like. *Primelephas korotorensis* is the only other species to be assigned to the genus.

Source (edited): "http://en.wikipedia.org/wiki/Primelephas"

Priscileo

Priscileo ("Ancient Lion") was the basal-most genus of Thylacoleonidae. Fossils of this genus have been found in Oligocene and Miocene strata of South Australia and Queensland. It was about the size of a possum and was less powerful than the other marsupial lions. The two species have crushing molar teeth (as in omnivorous kangaroos) as well as carnassial blades (large, flesh cutting teeth derived from the first lower molar and last upper molar). The Early Miocene species of *P. roskellyae* may have been an arboreal species.

Source (edited): "http://en.wikipedia.org/wiki/Priscileo"

Priscocamelus

Priscocamelus is an extinct genus of terrestrial herbivore the family Camelidae, endemic to North America from the Oligocene through Miocene 24.8—20.6 mya, existing for approximately 4.2 million years.

Taxonomy

Priscocamelus was named by Stevens (1969) [in Stevens et al.]. Its type is Priscocamelus wilsoni. It was assigned to Camelidae by Stevens (1969) and Carroll (1988).

Morphology

Body mass

Four specimens were examined for estimated body mass by M. Mendoza, C. M. Janis, and P. Palmqvist. These specimens were estimated to weigh:

- 106.3 kg (230 lb)
- 98.3 kg (220 lb)
- 91.4 kg (200 lb)
- 72.3 kg (160 lb)

Fossil distribution

Fossil distribution is restricted to two sites in the Rio Grande River valley of Texas.

Source (edited): "http://en.wikipedia.org/wiki/Priscocamelus"

Proailurus

Proailurnus was a prehistoric carnivore that lived in Europe and Asia approximately 25 million years ago in the Late Oligocene and Miocene. One recent phylogeny places it as a basal member of the Feloidea, the superfamily that includes mongooses, civets, hyenas, and cats; but other studies suggest that it instead was a felid (a true cat).

Proailurnus was a compact and small animal, just a little larger than the domestic cat, weighing about 20 lb (9 kg). It had a long tail, large eyes and sharp claws and teeth, with similar proportions to the modern viverrids. Its claws would have been retractable to some extent. Like the viverrids, *Proailurnus* was at least partially arboreal.

Proailurnus, a likely ancestor of *Pseudaelurus* which lived 20-10 million years ago, probably gave rise to the major felid lines, including the extinct machairodontines and the extant felines and pantherines, although the phylogeny of the cats is still not precisely known.

Source (edited): "http://en.wikipedia.org/wiki/Proailurus"

Procamelus

Procamelus is an extinct genus of terrestrial herbivore the family Camelidae, endemic to North America from the Oligocene through Miocene 20.6—4.9 mya, existing for approximately 15.7 million years.

Taxonomy

Priscocamelus was named by Leidy (1858). It is not extant. It was assigned to Camelidae by Leidy (1858) and Carroll (1988). The name is derived from the Greek πρό, meaning "before", and κάμελος (camel), thus "early camel."

Aepycamelus giraffinus and *Procamelus grandis*

Morphology

It had long legs designed for speed, and was about 1.3 metres (4.3 ft) in height, slightly smaller than a modern llama. Unlike modern camelids, it had a pair of small incisor teeth in the upper jaw. The remaining teeth were large and adapted for eating tough vegetation. The shape of the toes suggests that it possessed foot pads, like modern camels, but unlike earlier forms of camelid, which generally had hooves. This would have helped it walk over relatively soft ground.

Body mass

A single specimen was examined for estimated body mass by M. Mendoza, C. M. Janis, and P. Palmqvist. This specimen was estimated to weigh:

- 532.7 kg (1,200 lb)

It had a straighter neck than *Oxydactylus* or*Aepycamelus*.

Source (edited): "http://en.wikipedia.org/wiki/Procamelus"

Proconsul (primate)

Proconsul is an extinct genus of primates that existed from 23 to 5 million years ago during the Early Miocene epoch. Fossil remains are present in Eastern Africa including Kenya and Uganda. Four species have been classified to date: *P. africanus, P. heseloni, P. major* and *P. nyanzae*. The four species differ mainly in body size. Environmental reconstructions for the Early Miocene *Proconsul* sites are still tentative and range from forested environments to more open, arid grasslands.

They had a mixture of Old World monkey and ape characteristics, so their placement in the ape superfamily Hominoidea is tentative; some scientists place *Proconsul* outside of Hominoidea, before the split of the apes and Old World monkeys.

Proconsul's monkey-like features include pronograde postures, indicated by a long flexible back, curved metacarpals, and an above-branch arboreal quadrupedal positional repertoire. The primary feature linking *Proconsul* with extant apes is its lack of a tail; other "ape-like" features include its enhanced grasping capabilities, stabilized elbow joint and facial structure. *Proconsul* was definitely not suspensory like modern apes.

Discovery and classification

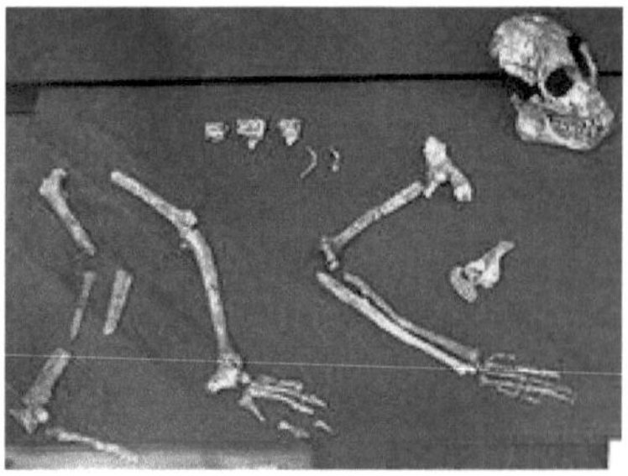

Proconsul nyanzae fossil, Muséum national d'histoire naturelle, Paris.

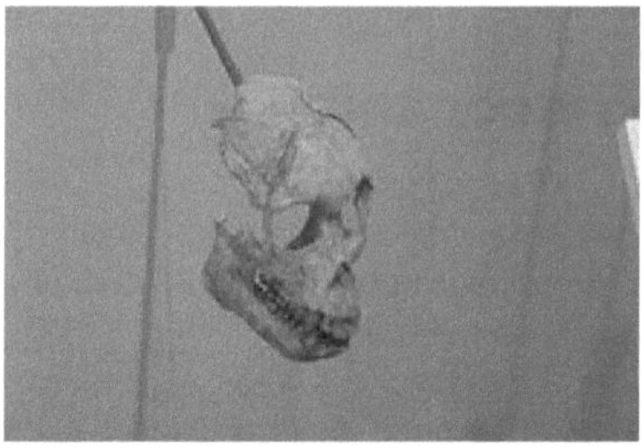

Proconsul heseloni skull.

The first specimen, a partial jaw discovered in 1909 by a gold prospector at Koru, near Kisumu in western Kenya, was also the oldest fossil hominoid known until recently, and the first fossil mammal ever found in sub-Saharan Africa. The name, Proconsul, was devised by Arthur Hopwood in 1933 and means "before Consul". At the time Consul was being used as a circus name for performing chimpanzees. The Folies Bergère of 1903 in Paris had a popular performing chimpanzee named Consul, and so did the Belle Vue Zoo in Manchester, England, in 1894. On the latter's death in that year Ben Brierley wrote a commemorative poem wondering where the "Missing Link" between chimpanzees and men was.

Hopwood in 1931 had discovered the fossils of three individuals while expeditioning with Louis Leakey in the vicinity of Lake Victoria. The Consul that he selected to use in the name was neither of the ones mentioned above, but another located in the London Zoo. Consul is being used Linnaean-style to symbolize the chimpanzee. Proconsul is therefore "ancestral to the Chimpanzee" in Hopwood's words. He also added africanus as the species name. It was the only one known at the time.

Other fossils discovered later were initially classified as africanus and subsequently reclassified; that is, the total pool of fossils originally considered *africanus* was "split" and the fragments "lumped" with other finds to create a new species. For example, Mary Leakey's famous find of 1948 began as *africanus* and was split from it to be lumped with Thomas Whitworth's finds of 1951 as *heseloni* by Alan Walker in 1993. This process creates some confusion for the public, which is told that *africanus* became *heseloni*. The finds from Koru and Songhor are still considered *africanus*. Four species are still defined even though many fossils have jumped species.

Restoration

The family of **Proconsulidae** was first proposed by Louis Leakey in 1962, eleven years after he and Wilfrid Le Gros Clark had defined *africanus*, *nyanzae* and *major*. It was not immediately accepted but ultimately prevailed.

The history of hominoid classification in the second half of the 20th century is sufficiently complex to warrant a few books itself. Most of the palaeoanthropologists have changed their minds at least once as new fossils have come to light and new observations have made, and will probably continue to do so. The classifications found in the literature of one decade are not generally the same as those of another. For example, in 1987 Peter Andrews and Lawrence Martin, established palaeontologists, took the point of view that Proconsul is not a Hominoid, but is a sister taxon to it.

Source (edited): "http://en.wikipedia.org/wiki/Proconsul_(primate)"

Proconsul africanus

Proconsul africanus is the first species of the Miocene-era fossil genus of primate to be discovered and was named by Arthur Hopwood, an associate of Louis Leakey, in 1933.

Discovery

The Leakey expedition of 1947-1948 to Rusinga Island in Lake Victoria uncovered more species of Proconsul. Mary Leakey made an especially complete find of *Proconsul* there in 1948, which was for a number of decades labeled *africanus*, but was reclassified as *heseloni* in 1993 by Alan Walker. In 1951 Leakey and Le Gros Clark placed Hopwood's **Xenopithecus koruensis** ("strange ape from Koru, Kenya") with africanus. In 1951 also T. Whitworth found more Proconsul on Rusinga, which he considered africanus, but they were lumped with 1992 finds by Walker to form heseloni.

The 18-million-year-old fossil species has been considered a possible ancestor of both great and lesser apes, and of humans. The paleontologist Louis Leakey, who was one of the foremost fossil-hunters of the 20th century and a champion of evolution, said:

> " An especially important creature was *Proconsul africanus*. This, many authorities once concluded, gave us an indication of the common stock for apes and men. We have good forelimb bones for it, and in 1948 on Rusinga Island Mary [Leakey] discovered a skull, the first nearly complete specimen ever found. Its canine teeth suggest an ape's, while its forehead reminds us of our own. It seems to me, however, to be neither an ancestral ape, nor yet an ancestor of man, but a side branch with characteristics of both stocks..." Leakey changed his mind a few times about the exact classification of Proconsul, as did most other palaeontologists. Opinion currently favors a position between the monkeys and the apes. "

Morphology

Proconsul africanus had a dental formula of 2:1:2:3 on both the upper and lower jaws. The molars of this species had thin enamel and there was a prominent molar cingula. This species also possessed a robust zygomatic bone and a pronounced snout. This species had a broad interorbital region and small frontoethmoidal sinuses. The maxillary sinus was restricted. This species had an auditory region which would be similar to that of extant apes and cercopithecoid monkeys. The ectotympanic tube was well-developed. This species lacked a tail and the canines of this species were sexually dimorphic. The skull lacks supraorbital tori and can be considered somewhat prognathous. This species has a cranial capacity of 167 cc and an encephalization quotient of 1.5. Based on the cranium, this species had an external brain surface much like that of gibbons and cercopithecoid monkeys. The wrist of this species has been described as monkey-like. This species has a talus in which the trochlear surface is highly curved and deeply grooved. The foot of this species possessed a transverse arch. *Proconsul africanus* had a brachial index of 96 which is comparable to the extant genus *Pan*. Overall the skeleton of this species can be described as being robust. This species had an average body mass of around 18 kilograms.

Diet

Based upon the dental morphology, it is conjectured that *Proconsul africanus* was a frugivorous species. A study by Aiello in 1981 concluded it was a "below-branch feeder."

Range and Locomotion

Based on postcranial pieces, *Proconsul africanus* was likely an arboreal quadruped.
Source (edited): "http://en.wikipedia.org/wiki/Proconsul_africanus"

Proconsul major

Proconsul major, an extinct primate of the genus *Proconsul*, was possibly the ancestor of *Afropithecus* and showed hominid characteristics. It occurred during the early Miocene and was roughly, the size of a gorilla.

Proconsul major lived on the continent of Africa. Based upon dental morphology, *Proconsul major* was a frugivorous species.

Morphology

Proconsul major had a dental formula of 2:1:2:3 on both the upper and lower jaw. The canines are sexually dimorphic. The inferior transverse torus is absent and the superior transverse torus is well-developed in *Proconsul major*. This species had an average body mass of around 50.0 kilograms.
Source (edited): "http://en.wikipedia.org/wiki/Proconsul_major"

Proconsul nyanzae

Proconsul nyanzae is a species of fossil primate first discovered by Louis Leakey on Rusinga Island in 1942, which he published in *Nature* in 1943. A joint publication of Wilfrid Le Gros Clark and Louis Leakey in 1951, "The Miocene Hominoidea of East Africa", first defines *Proconsul nyanzae*. In 1965 Simons and Pilbeam replaced *Proconsul* with *Dryopithecus*, using the same species names.

In 1967, Louis defined *Kenyapithecus africanus* on seven fossils from Rusinga Island. He saw it as an ancestor of *wickeri* and also of man, with a date of 20 mya in the middle Miocene. Another fossil found by the VanCouverings on Rusinga in 1967 seemed to confirm *africanus*. In 1969 Simons and Pilbeam moved *Kenyapithecus africanus* into *Dryopithecus nyanzae*. By 1978 the genus had recovered from the Dryopithecine event and was back to Proconsul. In that year Andrews moved Clark & Leakey's 1951 *Sivapithecus africanus* into *Proconsul nyanzae*.

A more recent discovery by Ward *et al.* in 1999 and reclassification splits *Kenyapithecus africanus* away again and lumps it with *Equatorius africanus*, which would move it to the Afropithecinae subfamily with *Afropithecus turkanensis*. As *Proconsul*, *Kenyapithecus* may not be in the same clade as apes and humans, but as the older *Equatorius*, it may be.

Morphology

Proconsul nyanzae had a dental formula of 2:1:2:3 on both the upper and lower jaw. The upper premolars of *Proconsul nyanzae* were large. This species had a relatively thick enamel on the molars. The mandible of this species was relatively robust. *Proconsul nyanzae* had an average body mass of about 30.0 kilograms.

Range

Proconsul nyanzae lived on the continent of Africa and the fossils were found in areas that suggest it lived in a dry, open woodland environment.
Source (edited): "http://en.wikipedia.org/wiki/Proconsul_nyanzae"

Proconsulidae

The **Proconsulidae** are an early family of primates that lived during the Miocene epoch in Kenya, and was restricted to Africa. The family is split into two subfamilies.

Members of the family have a mixture of Old World monkey and ape characteristics, so the placement in the ape superfamily Hominoidea is tentative; some scientists place Proconsulidae outside of Hominoidea, before the split of the apes and Old World monkeys. If Proconsulidae is not an ape family, then they would be somewhat closely related to *Aegyptopithecus*.

Discovery and classification

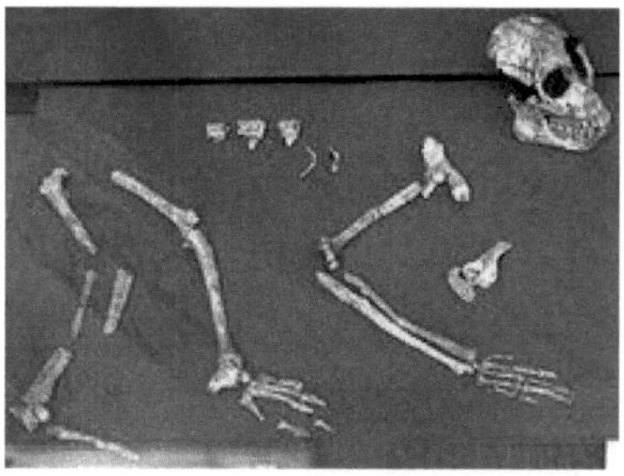

Proconsul nyanzae fossil, Muséum national d'histoire naturelle, Paris

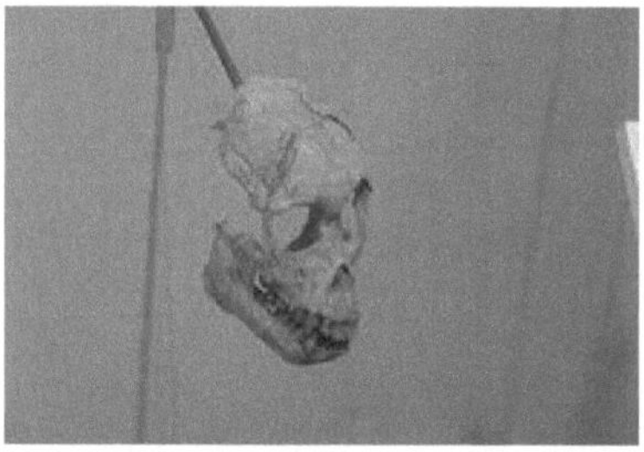

Proconsul heseloni skull

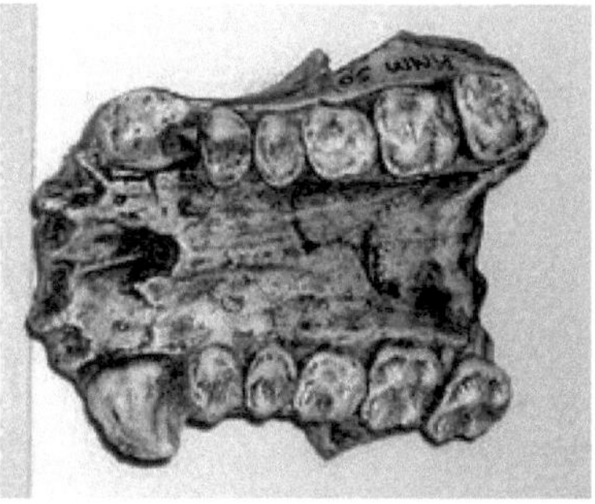

Rangwapithecus gordoni jaw

The first specimen, a partial jaw discovered in 1909 by a gold prospector at Koru, near Kisumu in western Kenya, was also the oldest fossil hominoid known until recently, and the first fossil mammal ever found in sub-Saharan Africa. The name, Proconsul, was devised by Arthur Hopwood in 1933 and means "before Consul." At the time Consul was being used as a circus name for performing chimpanzees. The Folies Bergère of 1903 in Paris had a popular performing chimpanzee named Consul, and so did the Belle Vue Zoo in Manchester, England, in 1894. On the latter's death in that year Ben Brierley wrote a commemorative poem wondering where the "Missing Link" between chimpanzees and men was.

Hopwood in 1931 had discovered the fossils of three individuals while expeditioning with Louis Leakey in the vicinity of Lake Victoria. The Consul that he selected to use in the name was neither of the ones mentioned above, but another located in the London Zoo. Consul is being used Linnaean-style to symbolize the chimpanzee. Proconsul is therefore "ancestral to the Chimpanzee" in Hopwood's words. He also added africanus as the species name. It was the only one known at the time.

Other fossils discovered later were initially classified as africanus and subsequently reclassified; that is, the total pool of fossils originally considered *africanus* was "split" and the fragments "lumped" with other finds to create a new species. For example, Mary Leakey's famous find of 1948 began as *africanus* and was split from it to be lumped with Thomas Whitworth's finds of 1951 as *heseloni* by Alan Walker in 1993. This process creates some confusion for the public, which is told that *africanus* became *heseloni*. The finds from Koru and Songhor are still considered *africanus*. Four species are still defined even though many fossils have jumped species.

The family of Proconsulidae was first proposed by Louis Leakey, eleven years after he and Wilfrid Le Gros Clark had defined *africanus*, *nyanzae* and *major*. It was not immediately accepted but ultimately prevailed.

The history of hominoid classification in the second half of the 20th century is sufficiently complex to warrant a few books itself. Most of the palaeoanthropologists have changed their minds at least once as new fossils have come to light and new observations have made, and will probably continue to do so. The classifications found in the literature of one decade are not generally the same as those of another. For example, in 1987 Peter Andrews and Lawrence Martin, established palaeontologists, took the point of view that Proconsul is not a Hominoid, but is a sister taxon to it.

Taxonomy

- Proconsuloidea or Hominoidea
  - **Proconsulidae** Leakey, 1963
    - Proconsulinae Leakey, 1963
      - *Proconsul* Hopwood, 1933
        - *Proconsul africanus* Hopwood, 1933
        - *Proconsul heseloni* Walker *et al.*, 1993
        - *Proconsul major* Le Gros Clark & Leakey, 1950
        - *Proconsul nyanzae* Le Gros Clark & Leakey, 1950
    - Nyanzapithecinae Harrison, 2002
      - *Nyanzapithecus'* Harrison, 1986
        - *Nyanzapithecus harrisoni* Kunimatsu, 1997
        - *Nyanzapithecus pickfordi* Harrison, 1986
        - *Nyanzapithecus vancouveringorum* Andrews, 1974
      - *Mabokopithecus* von Koenigswald, 1969
        - *Mabokopithecus clarki* von Koenigswald, 1969
      - *Rangwapithecus* Andrews, 1974
        - *Rangwapithecus gordoni* Andrews, 1974
      - *Turkanapithecus* Leakey & Leakey, 1986
        - *Turkanapithecus kalakolensis* Leakey & Leakey, 1986

Source (edited): "http://en.wikipedia.org/wiki/Proconsulidae"

Procranioceras

Procranioceras is an extinct genus of Artiodactyla, of the family Palaeomerycidae, endemic to Europe and Asia from the Miocene epoch, 16.3—13.6 Ma, existing for approximately 2.7 million years.

Taxonomy

Cranioceras (Procranioceras) was named by Frick (1937). It is not extant. Its type is *Procranioceras skinneri*. It was reranked as Procranioceras by Janis and Manning (1998) and Prothero and Liter (2007). It was assigned to Cranioceras by Frick (1937); to Dromomerycidae by Janis and Manning (1998); and to Cranioceratini by Prothero and Liter (2007)

Body mass

A single specimen of *Procranioceras* were measured by M. Mendoza, C. M. Janis, and P. Palmqvist for body mass.

- Specimen 1: 250.9 kg (550 lb)

Fossil distribution

- Bradley Site, Bone Valley Formation, Polk County, Florida
- Four Courners Site, Saskatchewan
- Hottell Ranch Horse Quarry, Banner County, Nebraska

Source (edited): "http://en.wikipedia.org/wiki/Procranioceras"

Prodeinotherium

Prodeinotherium (pro - "before" "terrible beast") is an early representative of the family Deinotheriidae, that lived in Africa, Europe, and Asia in the early and middle Miocene. It was the size of a small elephant, about 9 feet at the shoulders, but differed from elephants in possessing a pair of downward curving tusks on the lower jaw. In appearance and many characters it was like *Deinotherium* (with which it is placed in the subfamily, Deinotheriinae (Sanders *et al.* 2004), but differed in being of smaller size, having shorter forelimbs, and also in various details in the shape and form of the teeth.

The earliest species to appear is *Prodeinotherium hobleyi*, known from the Early Miocene of Kenya and Uganda (about 18 to 20 million years old). A molar of a small prodeinothere from Eritrea may also belong to this species and be of the same age.

Alternate view of skeleton

Prodeinotherium was larger and more specialised than its Oligocene predecessor *Chilgatherium*. It flourished for several millions of years, before being suddenly replaced in the middle Miocene by the much larger *Deinotherium*.

Source (edited): "http://en.wikipedia.org/wiki/Prodeinotherium"

Prolibytherium

Prolibytherium ("Before Libya's Beast") is an extinct artiodactyl ungulate native to Early Miocene North Africa.

The 1.80 m (6 ft) long creature was related to the modern giraffe and okapi. Unlike these, however, *Prolibytherium* had a set of large, leaf-shaped ossicones with a width of 35 cm (1 ft 2 in). These were probably used for display and fights, much like the antlers of a deer.

It was once regarded as a giraffid. Specifically, it was once regarded as the ancestral form of the African species of *Sivatherium* (*S. maurusium*), when it was originally described as "*Libytherium*." *Prolibytherium* is now regarded as a climacoceratid, due to the formation of its wide, palmate ossicones.

Source (edited): "http://en.wikipedia.org/wiki/Prolibytherium"

Promegatherium

Promegatherium ("Before *Megatherium*") is an genus of prehistoric xenarthrans that lived in South America, primarily Argentina, during the Middle to Upper Miocene. This genus is regarded as closely related to the later, and more famous genus, *Megatherium*, hence the reference in the name. The first specimens of *Promegatherium* were originally described by the biologist Florentino Ameghino in 1887. Its known species are *Promegatherium cabreri*, *P. nanum*, *P. parvulum*, *P. remulsum*, and *P. smaltatum*. *P. nanum* was originally placed in the genus

"*Eomegatherium*" before it was grouped together in *Promegatherium*.

Source (edited): "http://en.wikipedia.org/wiki/Promegatherium"

Prosantorhinus

Prosantorhinus is an extinct genus of rhinoceros from the lower and middle Miocene. The small teleoceratine rhinocerotid was found in western Europe and Asia. This group of creatures was a bit Hippopotamus like in form.

Source (edited): "http://en.wikipedia.org/wiki/Prosantorhinus"

Prosqualodon

Prosqualodon is an extinct genus of cetacean.

Prosqualodon was related to and looked like modern toothed whales. It was about 2.3 metres (7.5 ft) long and resembled a dolphin. It had long jaws with interlocking teeth that jutted to the outside, remaining visible when the jaws were closed, like those of a gharial.

In the back of the mouth it had triangular teeth similar to those of earlier cetaceans, but in most other respects, it was relatively advanced. It had developed the body form of modern whales, with a short neck and simple jaw structure, and like modern cetaceans, it also had a blowhole. The olfactory apparatus was reduced compared with earlier forms, suggesting that it had already lost much of its sense of smell, presumably relying on sound to catch its prey.

Source (edited): "http://en.wikipedia.org/wiki/Prosqualodon"

Prosynthetoceras

Prosynthetoceras is an extinct genus of Artiodactyla, of the family Protoceratidae, endemic to North America from the early Miocene epoch (Hemingfordian to Barstovian stages) 20.6—13.6 Ma, existing for approximately 7 million years.

Taxonomy

Prosynthetoceras was named by Patton (1967) [named as subgenus of Synthetoceras by Frick 1937; raised to genus level (?)]. It is not extant. It was assigned to Protoceratidae by Patton (1967) and Carroll (1988); and to Synthetoceratini by Webb (1981), Prothero (1998), Webb et al. (2003) and Prothero and Ludtke (2007).

Morphology

In appearance, *Prosynthetoceras* looked much like Syndyoceras, Kyptoceras, and Synthetoceras with rostral or snout horns and cranial horns above the orbital sockets.

Body mass

Four fossil specimens of *Prosynthetoceras* were measured by M. Mendoza, C. M. Janis, and P. Palmqvist for body mass.

- Specimen 1: 95.5 kg (210 lb)
- Specimen 2: 88.2 kg (190 lb)
- Specimen 3: 108.2 kg (240 lb)
- Specimen 4: 78.9 kg (170 lb)

Source (edited): "http://en.wikipedia.org/wiki/Prosynthetoceras"

Protoceras

Protoceras ('first horns') is an extinct genus of Artiodactyla, of the family Protoceratidae, endemic to North America from the Oligocene through Miocene (Orellan to Harrisonian stage) 33.9—20.6 Ma, existing for approximately 13.3 million years.

Taxonomy

Protoceras was named by Marsh (1891). Its type is ***Protoceras celer***. It is the genus of Protoceratidae and subfamily Protoceratinae. It was assigned to Protoceratidae by Marsh (1891) and Carroll (1988); and to Protoceratinae by Matthew (1908), Webb (1981), Prothero (1998), Webb et al. (2003) and Prothero and Ludtke (2007).

Morphology

Protoceras was 1 m (3 ft 4 in) long and resembled a deer in terms of body shape. Like some other protoceratids it had three pairs of blunt horns on its skull. In life these were probably covered with skin, much like the ossicones of a giraffe. *Protoceras* was sexually dimorphic - females only had one pair of horns (the one on the back of the skull), which was shorter than the same pair in males. Males probably used these horns for display, impressing females and/or intimidating rivals. Due to the orientation of the horns the males probably displayed them sideways instead of frontally.

Body mass

Three fossil specimens of *Poabromylus* were measured by M. Mendoza, C. M. Janis, and P. Palmqvist for body mass. The specimens were determined to weigh:

- Specimen 1: 95.2 kg (210 lb)
- Specimen 2: 57.2 kg (130 lb)
- Specimen 3: 89.8 kg (200 lb)
- Specimen 4: 54.6 kg (120 lb)

Protoceras was one of the earliest and most primitive protoceratids, still possessing upper incisors and four functional toes (later genera had only two functional, hooved toes). Protoceras lived in the deserts of the Late Oligocene, alongside the oreodont Leptauchenia.

Restoration of female and male by Charles R. Knight

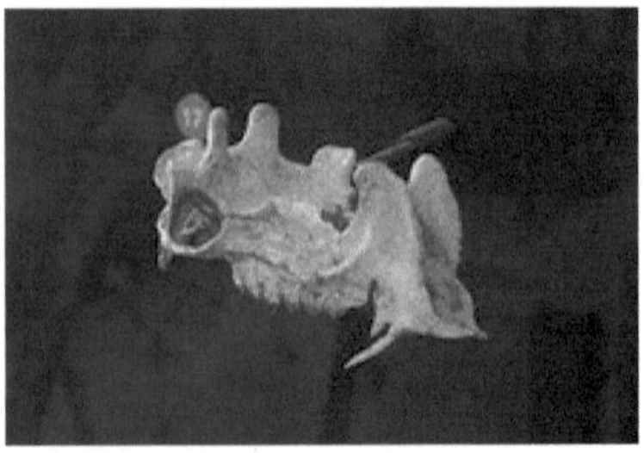

Male *Protoceras celer* skull

Restoration by Heinrich Harder

Source (edited): "http://en.wikipedia.org/wiki/Protoceras"

Protoceratidae

Protoceratidae is an extinct family of herbivorous North American artiodactyls (even-toed ungulates) that lived during the Eocene through Pliocene at around 46.2—4.9 Ma., existing for approximately 41.3 million years.

Taxonomy

Protoceratidae was named by Marsh (1891). Its type is *Protoceras*. It was assigned to Pecora by Cook in 1934; to Ruminantia by Thurmond and Jones (1981) and Spaulding et al. (2009); to Tylopoda by Carroll and by Webb et al in 2003. It was assigned to Artiodactyla by Marsh (1891), Hulbert and Whitmore (2006) and Prothero and Ludtke (2007). It was assigned to the infraorder Tragulina by Spaulding et al., 2009.

Morphology

Physically, they resembled deer, though not directly relating to them. They ranged from 1 to 2 meters in length, from about the size of a Roe Deer to an Elk. Unlike many modern ungulates, they lacked a cannon bone in their legs. Their dentition was similar to that of modern deer and cattle, suggesting that they fed on tough grasses and similar foodstuffs, with a complex stomach similar to that of camels. It is believed that at least some forms lived in herds.

The most dramatic feature of the protoceratids, however, were the horns of the males. In addition to having horns in the more usual place, protoceratids had additional, rostral horns above their nose. These horns were either paired, as in *Syndyoceras*, or fused at the base, and branching into two near the tip, as in *Synthetoceras*. In life the horns were probably covered with skin, much like the ossicones of a giraffe. The females were either hornless, or had far smaller horns than the males, and were therefore probably used in sexual display or competition for mates. In later forms, the horns were large enough to have been used in sparring between males, much as with the antlers of some modern deer.

Genera by epoch

Eocene

- *Heteromeryx*
- *Leptoreodon*
- *Leptotragulus*
- *Poabromylus*
- *Toromeryx*
- *Trigenicus*

Miocene

- *Paratoceras*
- *Protoceras*
- *Kyptoceras*
- *Lambdoceras*
- *Prosynthetoceras*
- *Pseudoprotoceras*
- *Synthetoceras*
- *Syndyoceras*

Source (edited): "http://en.wikipedia.org/wiki/Protoceratidae"

Protoceratinae

Protoceratinae is a subfamily of Protoceratidae belonging to the order Artiodactyla endemic to North America living during the Oligocene to Miocene epoch, 33.9—15.97 Ma, existing for approximately 17.93 million years.

Taxonomy

Protoceratinae was considered paraphyletic by Prothero (1998) and Webb et al. (2003). It was assigned to Hypertragulidae by Matthew (1908); and to Protoceratidae by Webb (1981), Prothero (1998), Webb et al. (2003) and Prothero and Ludtke (2007).

Members

Paratoceras and Protoceras

Fossil distribution

Fossil evidence is exclusive to *Protoceras celar* at 33.9 Ma recovered from Saskatchewan.
Source (edited): "http://en.wikipedia.org/wiki/Protoceratinae"

Protohippus

Protohippus is an extinct genus of horse.

Source (edited): "http://en.wikipedia.org/wiki/Protohippus"

Protolabis

Protolabis is an extinct genus of terrestrial herbivore the family Camelidae, endemic to North America from the Oligocene through Miocene 30.8—10.3 mya, existing for approximately 20.5 million years.

Taxonomy

Protolabis was named by Cope (1876). It was assigned to Camelidae by Cope (1876) and Carroll (1988).

Morphology

Body mass

Four specimens were examined for estimated body mass by M. Mendoza, C. M. Janis, and P. Palmqvist. These specimens were estimated to weigh:

- 451.7 kg (1,000 lb)
- 231.4 kg (510 lb)
- 342.7 kg (760 lb)
- 511.5 kg (1,100 lb)

Fossil distribution

Fossil distribution is widespread from Nicaragua, Central America to Montana and throughout the western U.S.
Source (edited): "http://en.wikipedia.org/wiki/Protolabis"

Protomarctus

Protomarctus is an extinct member of the *Borophaginae* and a small terrestrial canine (Bear-dog) which inhabited most of North America during the Burdigalian stage of the Miocene epoch 20.4—15.9 Mya, existing for approximately 4.5 million years.

Taxonomy

Protomarctus was named by Wang et al. (1999). It was assigned to Borophagini by Wang et al. (1999). One species, *Tomarctus optatus*, is known and was named by Matthew in 1924. The first specimen was found in Thomson Quarry, Sheep Creek Formation, Nebraska in a Hemingfordian layer.

Morphology

Protomarctus were short-faced, heavy-jawed canines usually massive in size. It would have shared its early existence with a variety of bear dogs like bone crushing canidae *Aelurodontina*, *Tomarctus*, *Phlaocyonini*, and *Rhizocyon*.

Body mass

Fossil specimens of two individuals' body mass were examined by Legendre and Roth. The first specimen was estimated to weigh 11.8 kg (26 lb). The second specimen was estimated to weigh 11.1 kg (24 lb).

Sister genera

- *Cormocyon*
- *Desmocyon*
- *Euoplocyon*
- *Metatomarctus*
- *Microtomarctus*
- *Psalidocyon*
- *Tephrocyon*

Fossil range

Protomarctus specimens have been found in present day Nebraska, Colorado, as far west as California and as far southwsest as New Mexico.
Source (edited): "http://en.wikipedia.org/wiki/Protomarctus"

Protypotherium

Protypotherium is an extinct genus of mammal native to South America during the Miocene epoch. A number of closely related animals date back further, to the Paleocene.

Restoration of *Protypotherium australe* and *Stegotherium tesselatum*

Protypotherium was slightly larger than a rabbit, measuring about 40 centimetres (1.3 ft) in length. The body and legs as well as the tail of this animal were relatively long, while its neck was short. It probably resembled a rodent, possessing slender limbs with clawed feet. Its rat-like skull contained a set of 44 unspecialized teeth.

From the shape of its claws, *Protypotherium* would have been adept at digging and likely took over the burrows of other animals. It is hypothesized that, while *Protypotherium* was chiefly herbivorous, it probably also ate carrion.

Source (edited): "http://en.wikipedia.org/wiki/Protypotherium"

Psalidocyon

Psalidocyon is an extinct genus of the *Borophaginae* and a small terrestrial canine which inhabited most of North America during the Hemingfordian stage of the Early Miocene subepoch through the Barstovian stage of the Middle Miocene subepoch 20.6—13.6 Ma, existing for about 7 mya..

Morphology

Fossil specimens of two individuals' body mass were examined by Legendre and Roth. The first specimen was estimated to weigh 6.53 kg (14 lb). The second specimen was estimated to weigh 6.31 kg (14 lb).

Species

P. marianae was named by Wang and others in 1999.

Sister genera

Cormocyon, *Desmocyon*, *Euoplocyon*, *Metatomarctus*, *Microtomarctus*, *Protomarctus*, and *Tephrocyon*.

Fossil distribution

It was initially found on Skull Ridge, Tesuque, New Mexico in a Miocene terrestrial horizon. It is the type species of Psalidocyon.

Source (edited): "http://en.wikipedia.org/wiki/Psalidocyon"

Pseudaelurus

Pseudaelurus is a prehistoric cat that lived in Europe, Asia and North America in the Miocene approximately 20-8 million years ago. It is an ancestor of today's felines and pantherines as well as the extinct machairodont saber-tooths, and is a successor to *Proailurus*. It originated from Eurasia and was the first felid to reach North America, when it entered the continent at about 18.5 m.y ending a 'cat-gap' of 7 million years. The slender proportions of the animal, together with its short, viverrid-like legs, suggest that it may have been an agile climber of trees.

Species and distribution

Europe

The first and most primitive species *Pseudaelurus turnauensis* (=*Pseudaelurus transitorius*) from the early Miocene was about the size of a domestic cat and probably evolved directly from the Oligocene *Proailurus*. The European *Pseudaelurus turnauensis* gave rise to additional species. *Pseudaelurus lorteti* was lynx-sized and the even larger *Pseudaelurus quadridentatus* weighed about 30 kg and was approximately the size of a cougar. The latter showed a trend towards slightly enlarged upper canines, indicating that it gave rise to the later saber-toothed machairodontines. Unexpectedly, the early *Pseudaelurus turnauensis* survived until 8 million years ago, at which time it is recorded in Germany, whereas the more derived *Pseudaelurus lorteti* and *Pseudaelurus quadridentatus* died out 2 million years earlier. A fourth European species is described as *Pseudaelurus romieviensis*, smaller than *P. quadridentatus* but larger than *P. lorteti*.

Africa and Asia

P. turnauensis is also recorded from the early Miocene Africa (Saudi Arabia). *P. lorteti* is also recognized from the middle Miocene of Asia (Jiangsu, China), where two additional species, *Pseudaelurus guangheensis* from the middle Miocene of Gansu (China) and *Pseudaelurus cuspidatus* from the middle Miocene of Xinjiang (China) are reported.

North America

Pseudaelurus intrepidus from the Early to Late Barstovian of North America was a relatively large species, overlapping in size with the European *P. quadridentatus*. A second large North

American species from the Early and Late Barstovian was *Pseudaelurus marshi. Pseudaelurus stouti* (originally described as *Lynx stouti*) also from the Early and Late Barstovian was a very small and slender from of the genus. *Pseudaelurus aeluroides* with only a type specimen from Early Barstovian from Nebraska is not well studied, but may be conspecific with *P. marshi*. A species *Pseudaelurus validus*, from the late Hemingfordian of New Mexico was described in 2001. This species overlaps in size with *P. intrepidus*, *P. marshi* and *P. quadridentatus*. Recently (2003) an additional North American species *Pseudaelurus skinneri* from the Late Hemingfordian of Nebraska was described. This species is smaller than *P. validus* and overlaps in size with the Eurasian species *P. lorteti*, *P. cuspidatus* and *p. guangheensis*.

Source (edited): "http://en.wikipedia.org/wiki/Pseudaelurus"

Pseudarctos

Pseudarctos is a member of the extinct family Amphicyonidae of terrestrial carnivores belonging to the suborder Caniformia, and which inhabited Euroasia from the Early Miocene subepoch to the Miocene epoch 16.9—11.1 Ma, existing for approximately 5.8 million years.

Pseudarctos was named Schlosser in 1899 and was assigned to *Amphicyonidae* by Carroll (1988). It was the size of a large fox.

Species

P. bavaricus

Sister genera

Afrocyon, Amphicyanis, Arctamphicyon, Brachycyon, Cynodictis, Drassonax, Goupilictis, Guangxicyon, Haplocyon, Haplocyonoides, Haplocyonopsis, Harpagophagus, Ictiocyon, Pachycynodon, Paradaphoenus, Protemnocyon, Pseudamphicyon, Pseudocyonopsis, Sarcocyon, Symplectocyon, Vishnucyon.

Fossil distribution

Pseudarctos has been uncovered in Tung Gur China, Sandberg, Slovakia, and Malartic, a la ferme Larrieu, France.

Source (edited): "http://en.wikipedia.org/wiki/Pseudarctos"

Pseudocyon

Pseudocyon, also known as ***Amphicyonopsis***, is an extinct genus of terrestrial carnivores belonging to the suborder Caniformia, family of Amphicyonidae ("bear dog") and which inhabited Euroasia and North America from the Miocene epoch to the Late Miocene subepoch living 23.3—7.2 Mya. *Pseudocyon* existed for approximately 5.3 million years.

Pseudocyon was assigned to Amphicyoninae by Hunt in 1988 and to Amphicyonidae by Lartet (1851), Carroll (1988) and Pickford et al. in 2000. It is synonymous with *Amphicyonopsis*.

Fossil distribution

Specimens were located in Belomechetskaia Russian Federation, Santa Cruz, New Mexico, Pontigne and Malartic, a la ferme Larrieu, France, and Nebraska.

Species

- *P. sansaniensis*
- *P. steinheimensis*
- *P. styriacus*

Source (edited): "http://en.wikipedia.org/wiki/Pseudocyon"

Pseudolabis

Pseudolabis is an extinct genus of terrestrial herbivore the family Camelidae, endemic to North America from the Oligocene through Miocene 24.8—20.6 mya, existing for approximately 4.2 million years.

Taxonomy

Pseudolabis was named by Matthew (1904). It was assigned to Camelidae by Matthew (1904) and Carroll (1988).

Morphology

Body mass

Four specimens were examined for estimated body mass by M. Mendoza, C. M. Janis, and P. Palmqvist. These specimens were estimated to weigh:

- 105.3 kg (230 lb)
- 84.4 kg (190 lb)
- 77.9 kg (170 lb)
- 54.3 kg (120 lb)

Fossil distribution

Fossil distribution is restricted to two sites in Wyoming and Nebraska.

Source (edited): "http://en.wikipedia.org/wiki/Pseudolabis"

Ptolemaiida

Ptolemaiida (or **the ptolemaiids**) is a taxon of wolf-sized mammals presumed to be descended from pantolestid cimolestids that lived in Northern and Eastern Africa during the Paleogene. The oldest fossils are from the latest

Eocene strata of the Jebel Qatrani Formation, near the Fayum oasis in Egypt.

Some experts place Ptolemaiida, *incertae sedis*, in Afrotheria, on the basis of paleobiology, as the taxon was endemic to Paleocene Africa, and because of some similarities in skull anatomical features in common with those in aardvarks.

There have been much confusion about the origins and even identities of the ptolemaiids. The first specimen, a set of isolated molar teeth, of the type species, *Ptolemaia lyonsi*, was originally identified as being a primate, as they were flat and nearly identical to those of primates. Later, when the first skull was found, it was then thought to be a monstrous, wolf-sized shrew, as the skull had long canine fangs, and was very gracile. However, recently, there has been reconsideration of the ptolemaiids' diet, and possible behavior, as wear on the teeth suggest it crushed hard or abrasive food, and that the teeth had little or no shearing ability. Even so, some sources still refer to them as being gigantic, carnivorous shrews.
Source (edited): "http://en.wikipedia.org/wiki/Ptolemaiida"

Puijila

Puijila darwini is an extinct species of pinniped which lived during the Miocene epoch about 21 to 24 million years ago. Approximately a metre (three feet) in length, the animal possessed only minimal physical adaptations for swimming. Unlike modern pinnipeds, it did not have flippers and its overall form was otter-like, albeit more specialized; its skull and teeth are the features that most clearly indicate that it is a seal. It is considered to be the most primitive member of the seal family yet found. The genus name is an Inuktitut word for a young seal; the species name honours the English naturalist Charles Darwin. The one known specimen is a nearly complete fossilised skeleton. It is being housed at the Canadian Museum of Nature in Ottawa, Ontario.

Background

Skeleton

Puijila darwini was a semi-aquatic carnivore which represents a morphological link in early pinniped evolution. Its fossil remains demonstrate the presence of enlarged, probably webbed feet, robust forelimbs and an unspecialized tail. This suggests that *Puijila* swam quadrupedally using its webbed fore and hind feet for propulsion. Phylogenetic studies including molecular evidence suggest a sister relationship between pinnipeds (seals) and ursoids (bears) as well as musteloids (weasels and otters). It had been popularly assumed that land-dwelling mammals had at some point transitioned to a more marine existence, in essence "returning to the sea" in order to gain some sort of survival advantage. However, fossil evidence of this transition had been weak or contentious. The discovery of *Puijila* is important as it represents a morphological link in early pinniped evolution, and one that appears morphologically, to precede the more familiarly structured *Enaliarctos* genus, despite apparently being a younger genus. In other words, *Puijila* is a transitional fossil that provides information about how the seal family returned to the seas, similar to the way that *Archaeopteryx* illuminates the origin of modern birds.

Discovery

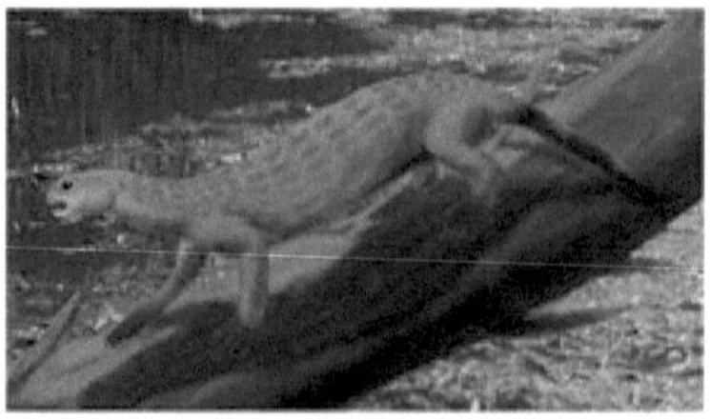

Life restoration of *Puijila darwini*.

This novel species was discovered in 2007 by Natalia Rybczynski and her team using surface collection and screening at an early Miocene lake deposit of the Haughton Formation of Devon Island in Nunavut, Canada. The palaeobotanical record suggests that the palaeoenvironment around the lake comprised a forest community transitional between a boreal and a conifer–hardwood forest, in a cool temperate, coastal climate with moderate winters. *Puijila darwini* is the first mammalian carnivore found in the Haughton lake deposits. This also gives an indication that the entire pinniped family may have originated in the Arctic.

The initial find is credited to field assistant Elizabeth Ross, and was partly a matter of luck. Ross had been unexpectedly stranded with the team's ATV which had run out of fuel several kilometers away from base camp. The brain case was discovered a year later on the first day of the 2008 field expedition by Martin Lipman, the team's photographer.
Source (edited): "http://en.wikipedia.org/wiki/Puijila"

Pyramios

Pyramios is an extinct genus of Diprotodontia from the Miocene of Australia.
Source (edited): "http://en.wikipedia.org/wiki/Pyramios"

Ramoceros

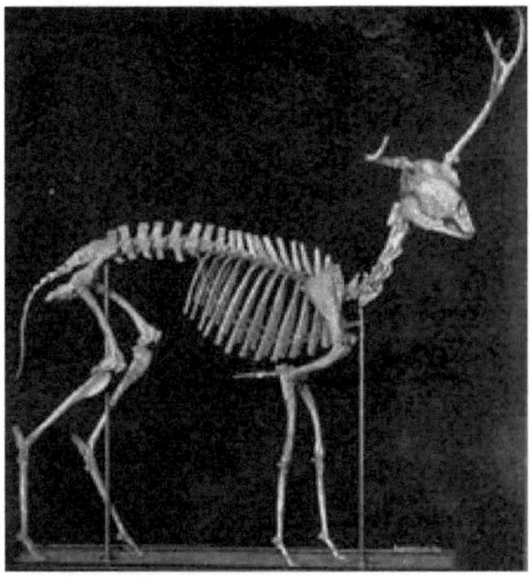

Ramoceros osborni

Ramoceros is an extinct genus of artiodactyl.
Source (edited): "http://en.wikipedia.org/wiki/Ramoceros"

Rhynchotherium

Rhynchotherium is an extinct genus of proboscidea endemic to North America and Central America during the Miocene through Pliocene from 13.650—3.6 Ma, living for approximately 10 million years.

This gomphothere had two tusks and may have evolved from *Gomphotherium*.

Taxonomy

Rhynchotherium was named by Falconer (1868). Its type is *Rhynchotherium tlascalae*. It was assigned to Gomphotheriidae by Falconer (1868), Carroll (1988), Lambert and Shoshani (1998) and Corona-M and Alberdi (2006).
Source (edited): "http://en.wikipedia.org/wiki/Rhynchotherium"

Riversleigh Marsupial Mole

The **Riversleigh Marsupial Mole** (*Notoryctes* sp.) is an extinct marsupial mole from the early Miocene.
Source (edited): "http://en.wikipedia.org/wiki/Riversleigh_Marsupial_Mole"

Rytiodus

Rytiodus is an extinct genus of sirenian, whose fossils have been discovered in Europe.

Description

With a length of 6 m (20 ft), *Rytiodus* was about twice the size as modern sirenians, surpassed only by Steller's sea cow, which was up to 8 m (27 ft) long. Like its closest modern relatives, the dugongs, *Rytiodus* had a pair of flippers, a streamlined body and a tail fin. Its flattened snout allowed it to feed in shallow coastal waters. Unlike modern sirenians, *Rytiodus* had short tusks which it may have used to extract food from the sand.
Source (edited): "http://en.wikipedia.org/wiki/Rytiodus"

Sahelanthropus

Sahelanthropus tchadensis is an extinct hominid species that is dated to about 7 million years ago. Whether it can be regarded as part of the Hominina tree is unclear; there are arguments both supporting and rejecting it. Another complication in its classification is that it is older than the human-chimpanzee divergence (estimated to 6.3 to 5.4 million years ago) seen in genetic data, and that there are few if any specimens other than the partial cranium known as *Toumaï*.

Fossils

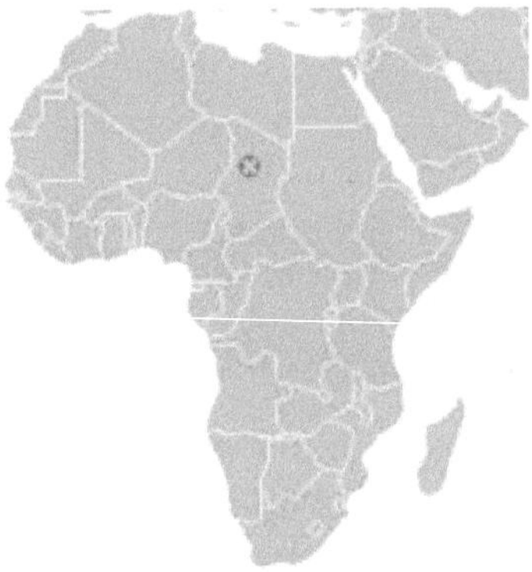

Location of discovery

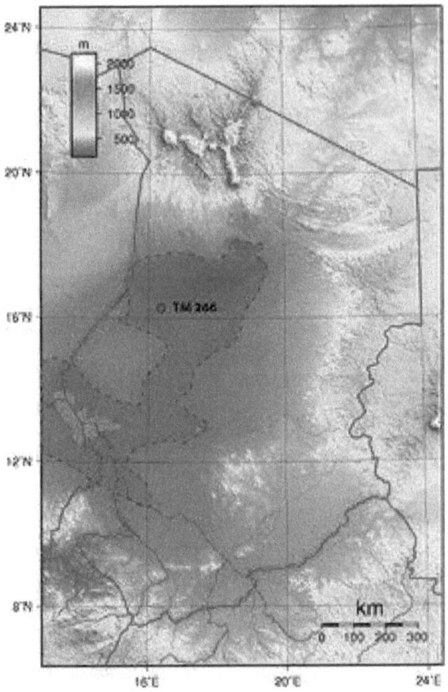

Detail of map

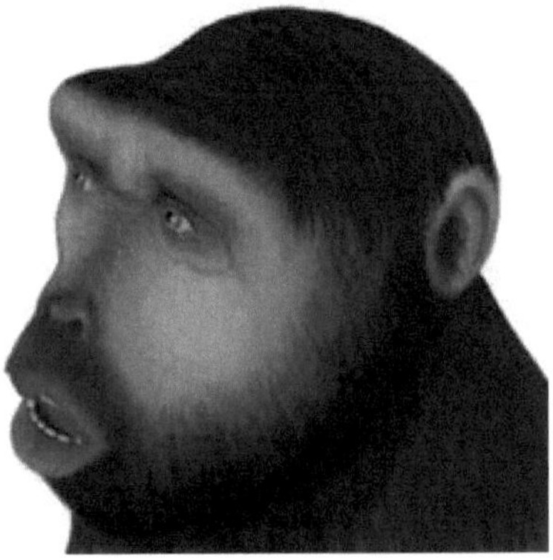

Restoration of the face of *Sahelanthropus*

Existing fossils – a relatively small cranium nicknamed **Toumaï** ("hope of life" in the local Dazaga language of Chad), five pieces of jaw and some teeth–make up a head that has a mixture of derived and primitive features. The braincase, being only 320 cm³ to 380 cm³ in volume is similar to that of extant chimpanzees and is notably less than the approximate human volume of 1350 cm³. The teeth, brow ridges, and facial structure differ markedly from those found in *Homo sapiens*. Cranial features show a flatter face, u-shaped dental arcade, small canines, an anterior foramen magnum, and heavy brow ridges. No postcranial remains have been recovered. Due to the distortion that the cranium has suffered, a 3D computer reconstruction has not been produced.

Since no postcranial remains (bones below the skull) have been discovered, it is as of yet unknown whether *Sahelanthropus tchadensis* was indeed bipedal, although claims for an anteriorly placed foramen magnum suggests that this may have been the case, some paleontologists have disputed this interpretation of the basicranium. Its canine wear is similar to other Miocene apes. Moreover, according to recent information, the femur of a hominid may have been discovered alongside the cranium but never published.

The fossils were discovered in the Djurab desert of Chad by a team of four led by Michel Brunet; three Chadians, Adoum Mahamat, Djimdoumalbaye Ahounta and Gongdibé Fanoné, and Frenchman, Alain Beauvilain. All known material of *Sahelanthropus* were found between July 2001 to March 2002 at three sites (TM 247, TM 266 which yielded most of the material, and TM 292). The discoverers claimed that *S. tchadensis* is the oldest known human ancestor after the split of the human line from that of chimpanzees. The bones were found far from most previous hominin fossil finds, which are from Eastern and Southern Africa. However, an *Australopithecus bahrelghazali* mandible was found in Chad by Beauvilain A., Brunet M. and Moutaye A.H.E. as early as 1995. With the sexual dimorphism known to have existed in early hominids, the difference between *Ardipithecus* and *Sahelanthropus* may not be large enough to warrant a separate species for the latter.

Relationship to modern humans and great apes

Sahelanthropus may represent a common ancestor of humans and chimpanzees; no consensus has been reached yet by the scientific community. The original placement of this species as a human ancestor but not a chimpanzee ancestor would complicate the picture of human phylogeny. In particular, if Toumaï is a direct human ancestor, then its facial features bring the status of *Australopithecus* into doubt because its thickened brow ridges were reported to be similar to those of some later fossil hominids (notably Homo erectus), whereas this morphology differs from that observed in all australopithecines, most fossil hominids and extant humans.

Another possibility is that Toumaï is related to both humans and chimpanzees, but is the ancestor of neither. Brigitte Senut and Martin Pickford, the discoverers of *Orrorin tugenensis*, suggested that the features of *S. tchadensis* are consistent with a female proto-gorilla. Even if this claim is upheld, then the find would lose none of its significance, for at present precious few chimpanzee or gorilla ancestors have been found anywhere in Africa. Thus if *S. tchadensis* is an ancestral relative of the chimpanzees (or gorillas) then it represents the first known member of *their* lineage. Furthermore, *S. tchadensis* does indicate that the last common ancestor of humans and chimpanzees is unlikely to resemble chimpanzees very much, as had been previously supposed by some paleontologists. · Unfortunately, the exact age of the fossil is somewhat hard to determine. While molecular clocks are increasingly found to be far more unreliable than initially believed sediment isotope analysis which yielded an age of about 7 million years is generally considered quite reliable. In this case however, the fossils were found exposed in loose sand; co-discoverer Beauvilain cautions that such sediment can be easily moved by the wind, unlike packed earth. In fact, Toumaï was probably reburied in the recent past. Taphonomic analysis reveals the likelihood of one, perhaps two, burial(s) which seemingly occurred after the introduction of Islam in the region. Two other hominid fossils (a left femur and a mandible) were in the same "grave" along with various mammal remains.The sediment surrounding the fossils might thus not be the material the bones were originally deposited in, making it necessary to corroborate the fossil's age by some other means. The fauna found at the site – namely the anthracotheriid *Libycosaurus petrochii* and the *suid Nyanzachoerus syrticus* – suggests an age of more than 6 million years, as these species were probably extinct already by that time.

Source (edited): "http://en.wikipedia.org/wiki/Sahelanthropus"

Samburupithecus

Samburupithecus was a primate that lived in Kenya during the middle to late Miocene. The one species in this genus, *Samburupithecus kiptalami*, is known only from a maxilla fragment dated to 9.5 Ma discovered by Hidemi Ishida and Martin Pickford in 1997. *Samburupithecus* was approximately 60 kg and was most likely a frugivorus terrestrial quadruped (Fleagle, 1999). Paleoenvironmental reconstructions indicate that *Samburupithecus* most likely lived in a wooded habitat surrounded by savannah (Tsujikawa, 2005).

Defining cranial traits of this genus include low, broad zygomatics, straight alveolar process and large maxillary sinus. Defining dental traits include three-rooted premolars, thick enamel and buonodont cusps (Ward, 2002). Unlike many other Miocene hominoids, *Samburupithecus* has teeth that are elongated from front to back. This feature links the taxon to gorillas, chimpanzees and hominins, but its relationships within this clade are presently unknown (Fleagle, 1999).

Source (edited): "http://en.wikipedia.org/wiki/Samburupithecus"

Samotherium

Samotherium is an extinct genus of giraffe from the Miocene and Pliocene of Eurasia and Africa. *Samotherium* had two ossicones on its head, and long legs. The ossicones usually pointed upward, and were curved backwards, with males having larger, more curved ossicones, though, in the Chinese species, *S. sinense*, the straight ossicones point laterally, not upwards. The genus is closely related to *Shansitherium*. The skull of a *Samotherium* is portrayed almost intact on an ancient Greek vase as a monster that Heracles is fighting.

Etymology

The genus name translates as "Beast of Samos," to commemorate where the first fossils were found.

Source (edited): "http://en.wikipedia.org/wiki/Samotherium"

Scalabrinitherium

Scalabrinitherium is an extinct genus of the family Macraucheniidae. Fossils of this animal were found amongst the fossils of prehistoric xenarthrans.

Source (edited): "http://en.wikipedia.org/wiki/Scalabrinitherium"

Scaphohippus

Scaphohippus is an extinct Miocene genus of equine, with two known species, known from fossils found in California, New Mexico, Montana, and Nebraska.

History

Both species in the genus were originally described as members of the wastebasket taxon *Merychippus* in the 1915 by John Merriam. The genus was described from specimens found in the Barstow Formation of southeast California. Specimens of *Scaphohippus* were medium-sized horses with hypsodont (high-crowned) teeth.

Phylogenetic relationships

The genus is closely related to *Protohippus* and *Callippus*. In the Barstow Formation, the two species of *Scaphohippus* occur with only one other hypsodont horse, *Acritohippus stylodontus*. *Scaphohippus* apparently developed on the Great Plains late in the Hemingfordian, about 17 to 16 million years ago and invaded the Great Basin at about 15 million years ago. The Great Plains and Great Basin were then isolated by tectonic barriers and *Scaphohippus* became extinct in the Great Plains, but survived in the Great Basin, until the barriers disappeared at about 13 million years ago and *Scaphohippus* also became extinct in the Great Basin.

Species

Scaphohippus sumani

Scaphohippus sumani, the type species for the genus, was originally described in 1915 by John Merriam as *Merychippus sumani* from University of California Museum of Paleontology specimen number 21422, consisting of four teeth. The specimen was collected from fossiliferous layers of the Barstow Formation outcropping in Rodent Hill Basin, San Bernardino County, California. Specimens of the species have also been found in three other sites in California and one in Nebraska. While numerous specimens of isolated teeth and partial palates have been found at the type locality, comparatively few complete skulls are known. The few examples from the location are badly crushed or have been altered due to pressure during fossilization. Detailed measurement of teeth from the type locality enabled identification of a *S. sumani* skull from the Punchbowl Formation, Cajon Valley, California and one from the Olcott Foramtion, Sioux County, Nebraska.

Scaphohippus intermontanus

Scaphohippus intermontanus, the second species in the *Scaphohippus* genus, is distinguishable from *S. sumani* by several distinct feature of the teeth. The upper cheek teeth have a height of 40–50 mm at the mesostyle and the protocone connects with the protoconule after approximately 30% wear. The enamel fossettes are both simpler and possess fewer plications than those of *S. sumani*. The species was first described as *Merychippus intermontanus* in the same 1915 paper by John Merriam in which *"Merychippus" sumani* was described. The type specimen for *S. intermontanus* is a section of jaw containing 6 teeth found in the *Hemicyon* Tuff, Barstow Formation, that outcrops in Hellgate Basin, San Bernardino County. The species has been confidently identified from several California sites with possible specimens known from Montana and New Mexico. Source (edited): "http://en.wikipedia.org/wiki/Scaphohippus"

Scelidotheriidae

Scelidotheriidae is a family of extinct mammals within the order of Pilosa and suborder Folivora. This family of ground sloths is related to the other families of extinct ground sloths, being the Megatheriidae, the Mylodontidae, the Nothrotheriidae, and the Orophodontidae. The only extant families of the suborder Folivora are the Bradypodidae and the Megalonychidae.

Together with the Mylodontidae and the engimatic *Pseudoprepotherium*, the scelidotheriids form the superfamily Mylodontoidea. *Chubutherium* is an ancestral and very plesiomorphic member of this family and does not belong to the main group of closely-related genera.

FAMILY †SCELIDOTHERIIDAE Ameghino, 1889

- Genus †*Chubutherium* (basal)
- Subfamily †Scelidotheriinae
    - Genus †*Scelidotheriops*
    - Genus †*Analcitherium*
    - Genus †*Nematherium*
    - Genus †*Neonematherium*
    - Genus †*Elassotherium*
    - Genus †*Scelidotherium*
    - Genus †*Catonyx*
    - Genus †*Proscelidodon*

Source (edited): "http://en.wikipedia.org/wiki/Scelidotheriidae"

Serbelodon

Serbelodon is an extinct genus of African and European Proboscidean. It was a gomphothere "shovel-tusker." It had tusks and a trunk. It lived during the Miocene Epoch, and may have evolved from *Amebelodon*.

In 1933, Serbelodon burnhami (now *Amebelodon burnhami*) was newly discovered and named after the celebrated scout Frederick Russell Burnham. Source (edited): "http://en.wikipedia. org/wiki/Serbelodon"

Shansitherium

Shansitherium ("Beast of Shanxi") is an extinct genus of superficially moose- or antelope-like giraffe from the late Miocene epoch of Shanxi Province, China. They were closely related to the genus *Samotherium*.

Mounted *S. tafeli* skeleton from the Beijing Museum

Source (edited): "http://en.wikipedia. org/wiki/Shansitherium"

Shennongtherium

Shennongtherium is an extinct genus of rhinoceros from the Miocene time period. It once roamed in what is now China.

Source (edited): "http://en.wikipedia. org/wiki/Shennongtherium"

Silvabestius

Silvabestius is an extinct genus of marsupial dating to the Early Miocene. They were grazing animals about the size of a modern sheep.

This animal is known from two skulls found close together which have come to be known as the "Madonna and Child" fossils.

Source (edited): "http://en.wikipedia. org/wiki/Silvabestius"

Simocyon

Simocyon ("short-snouted dog") is a genus of extinct mammal belonging to the order Carnivora. *Simocyon*, which was about the size of a mountain lion, lived in the late Miocene and early Pliocene epochs, and has been found in Europe, Asia, and, rarely, North America (Peigné et al., 2005) and Africa. The relationship of *Simocyon* to other carnivores has been controversial, but studies of the structure of its ear, teeth, and ankle now indicate that its closest living relative is the red panda, *Ailurus* (Wang, 1997; Peigné et al., 2005), although it is different enough to be classified in a separate subfamily (Simocyoninae) along with related genera *Alopecocyon* and *Actiocyon*. While the red panda is primarily herbivorous, the teeth and skull of *Simocyon* indicate that it was carnivorous, and it may have engaged in some bone crushing, like living hyaenas (Peigné et al., 2005). The skeleton of *Simocyon* indicates that, like the red panda, it could climb trees, although it probably also spent considerable time on the ground (Salesa et al., 2006). *Simocyon* and *Ailurus* both have a radial sesamoid, an unusual bone in the wrist that acts like a false thumb (Salesa et al., 2006). Its predators were tremarctine bears, nimravid cats, and early canids.

Source (edited): "http://en.wikipedia. org/wiki/Simocyon"

Sinodonomys

Sinodonomys is an extinct genus of dipodid rodent which existed in central Nei Mongol, China, during the early Miocene (middle Burdigalian age). It was first named by Yuri Kimura in 2010 and the type species is *Sinodonomys* *simplex.*

Source (edited): "http://en.wikipedia. org/wiki/Sinodonomys"

Sinomastodon

Sinomastodon ("Chinese mastodont") is an extinct gomphothere genus (of order Proboscidea), from the Late Miocene to the Early Pleistocene deposits of south-east Asia (China, Japan, and Indonesia). It is not to be confused with the genus *Mammut* from a different proboscidean family, whose members are commonly called "mastodons".

The animal was very similar to modern elephants with size ranging from 3.6 to 5.3 m. Several species are known from China, the best known being *S. hanjiangensis* from the Late Miocene and Early Pliocene of the Shanxi province. It is known from an incomplete skeleton of an adult (measuring 5.3 m) including maxilla, mandibles, teeth, tusks and other materials have been discovered. The Japanese species *S. sendaicus* described in 1924 from dentary materials from Pliocene deposits has been ascribed to the genus, as well as the species *S. bumiajuensis* (formerly *Tetralophodon*) from the late Pliocene of Java.

Source (edited): "http://en.wikipedia. org/wiki/Sinomastodon"

Sinotherium

Sinotherium ("Chinese Beast") was a genus of single-horned rhinoceri of the late Miocene and Pliocene. It was ancestral to *Elasmotherium*, and its fossils have been found in western China. *Sinotherium* diverged from the ancestral genus, *Iranotherium*, first found in Iran, during the early Pliocene. Some experts prefer to lump *Sinotherium*, and *Iranotherium* into *Elasmotherium*.

Its synonym, "*Paraelasmotherium*," is often used to refer to the Miocene specimens.

Source (edited): "http://en.wikipedia. org/wiki/Sinotherium"

Sivapithecus

Sivapithecus is a genus of extinct primates. Fossil remains of animals now assigned to this genus, dated from 12.5 million to 8.5 million years old in the Miocene, have been found since the 19th century in the Siwalik Hills in what is now India, Nepal, and Pakistan. Any one of the species in this genus may have been the ancestor to the modern orangutans.

Description

Sivapithecus was about 1.5 metres (4.9 ft) in body length, similar in size to a modern orangutan. In most respects, it would have resembled a chimpanzee, but its face was closer to that of an orangutan. The shape of its wrists and general body proportions suggest that it may have spent a significant amount of its time on the ground, as well as in trees. It had large canine teeth, and heavy molars, suggesting a diet of relatively tough food, such as seeds and savannah grasses.

Species

Currently three species are generally recognized. *Sivapithecus indicus* fossils date from about 12.5 million to 10.5 million years ago. *S. sivalensis* lived from 9.5 million to 8.5 million years ago. It was found at the Pothowar plateau in Pakistan as well as in parts of India.

The animal was about the size of a chimpanzee but had the facial morphology of an orangutan; it ate soft fruit (detected in the toothwear pattern) and was probably mainly arboreal.

In 1988 a third, significantly larger species was described and named *S. parvada* (dated at about 10 million years ago).

In 1982, David Pilbeam published a description of a significant fossil find — a large part of the face and jaw of a *Sivapithecus*. The specimen bore many similarities to the orangutan skull and strengthened the theory (previously suggested by others) that *Sivapithecus* was closely related to orangutans.

Ramapithecus

Siwalik specimens once assigned to the genus *Ramapithecus* are now considered by most researchers to belong to one or more species of *Sivapithecus*. *Ramapithecus* is no longer regarded as a likely ancestor of humans.

The first incomplete specimens of *Ramapithecus* were found in Nepal on the bank of Tinau River western part of the country in 1932. The finder (G. Edward Lewis) claimed that the jaw was more like a human's than any other fossil ape then known. In the 1960s this claim was revived. At that time, it was believed that the ancestors of humans had diverged from other apes 14 million years ago. Biochemical studies upset this view, suggesting that there was an early split between orangutan ancestors and the common ancestors of chimps, gorillas and humans.

Meanwhile, more complete specimens of *Ramapithecus* were found in 1975 and 1976, which showed that it was less human-like than had been thought. It began to look more and more like *Sivapithecus* - meaning that the older name must take priority. It could be that *Ramapithecus* was just the female form of *Sivapithecus*. They were definitely members of the same genus. It is also likely that they were already separate from the common ancestor of chimps, gorillas and humans, which may be represented by the prehistoric great ape, *Nakalipithecus nakayamai*.

Source (edited): "http://en.wikipedia.org/wiki/Sivapithecus"

Sprite Possum

The **Sprite Possum** (Ektopodontid) is an extinct genus of marsupial from the early Miocene.

Source (edited): "http://en.wikipedia.org/wiki/Sprite_Possum"

Squalodon

Squalodon

Squalodon is an extinct genus of whales, belonging to the family Squalodontidae. Named by Grateloup in 1840, it was originally believed to be an iguanodontid dinosaur but has since been reclassified. The name *Squalodon* comes from *Squalus*, a genus of shark. As a result its name means "Squalus tooth."

Description

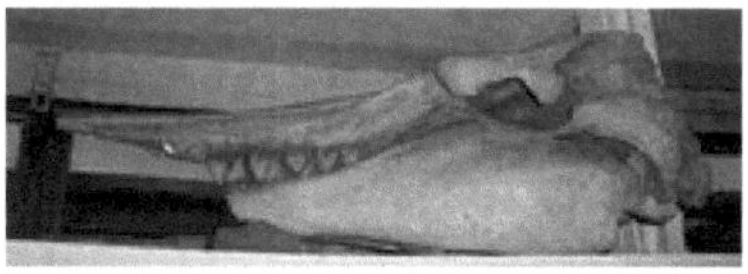

Squalodon bariensis skull

Squalodontids are odontocetes that lived during the early-middle Oligocene into the middle Miocene, about 33 to 14 million years ago. The genus *Squalodon* belongs to the order Odontoceti, the toothed whales (Kowinsky, J.). They are named after the shark *Squalus* because their cheek teeth look like the teeth of a squalus shark. The unique-looking squalodontids were likely distributed throughout the world in warm waters during the Oligocene and Miocene. Squalodontidae went extinct in the middle of the Miocene, leaving no descendants. Hypotheses of why this family lead to extinction have to deal with competition of other groups of dolphins as well as climate change (Fordyce, R.E).

Ancestral and modern features

These whales are characterized by both ancestral and modern features. Their teeth are the most evident ancestral feature. At this time in history other toothed whales were evolving simple conical teeth while Squalodontidae retained their primitive dentition that their ancestors (the archaeocetes) had developed (Fordyce, R.E.). Today living odontocetes have little variation in their teeth. Squalodontids' teeth are much more complex: they are widely spaced apart; their cheek teeth are triangular and serrated for grasping and cutting. Due to the efficiency of their primitive dentition squalodontids could have a diverse variety of prey (Fordyce, R.E.),(Kowinsky, J.). Another ancestral quality of the Squalodontidae is their necks. Squalodontid necks are more compressed than their ancestors, the Archaeoceti. Compared to toothed whales at that time, the squalodontids were likely more mobile. Paleontologists also believe that the dorsal fins were reduced but larger than that of the ancestors (Kowinsky, J.). Shark toothed whales also possess many modern features. Their crania were well compressed, their rostrums were telescoped outward, and their skulls show proof of the origin of echolocation. (Kowinsky, J.)

Fossil record and classification

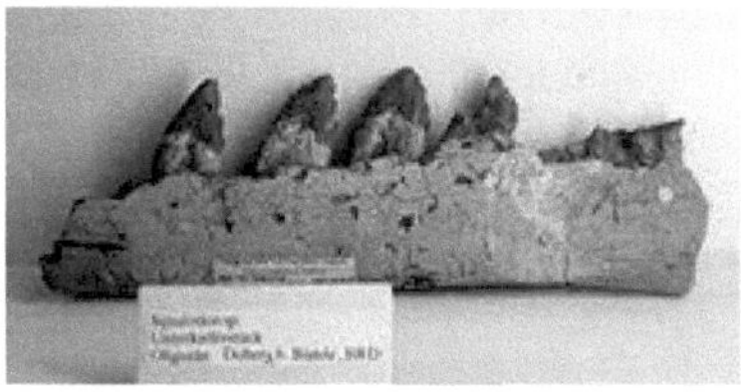

Part of the lower jaw of a *Squalofon sp.*, Doberg/Bünde, District of Herford, North Rhine-Westphalia, Germany. Museum für Naturkunde (Berlin).

One of the first fossils described in scientific literature is that of a specimen of *Squalodon* (Fordyce, R.E.). Fossils of this species are identified mainly by the teeth but also somewhat by their quite complete skull. Most of the fossil record consists of teeth. Some fossils are of fragments of the skull. A complete fossil skull has yet to be discovered. These odontocete fossils have been discovered in Europe, eastern North America, and Argentina. Because tooth fossils are insufficient for species identification, the fossil record only indicates the family (Dooley,A.). The fossils of squalodontids indicate that this species is more closely related to endangered species of dolphins and not to most of the living dolphins today (Fordyce, R.E.). The classification of this genus is not clear. A book by Charles Darwin states that "....*Squalodon*, which have been placed by some naturalists in an order by themselves, are considered......to be undoubtedly cetaceans..." (Darwin, C.) Many of the fresh-water dolphins are differentiated phylogenitically very well, while the argument of some of the species has been going on for more than a century. The taxon is characterized during the Oligocene and Miocene in which heterodont teeth are standard amongst the family. Some modern features of the scapula, however, contradict with current phylogenetic relationships. Squalodontids were believed to be the last common ancestor of the odontocetes until 1984. Muizon came to the conclusion that rather than to any of the living species this family is closer related to the endangered species. Therefore, the ancestry of today's dolphins has little to do with the squalodontids

(Fordyce, R.E.). Source (edited): "http://en.wikipedia. org/wiki/Squalodon"

Stegodontidae

Stegodontidae is an extinct family of proboscideans that lived from the Miocene through Pleistocene period endemic to Africa and Asia from 23.03 to 1.810 Ma, living for approximately 21. 22 million years.

Taxonomy

Stegodontidae was named by Osborn (1918). It was assigned to Mammutoidea by Carroll (1988); to Elephantoidea by Lambert and Shoshani (1998); and to Elephantoidea by Shoshani et al. (2006).

Source (edited): "http://en.wikipedia. org/wiki/Stegodontidae"

Stegolophodon

Tooth

Stegolophodon is an extinct genus of proboscidean. *Stegolophodon* was a gomphothere with four tusks and a trunk. It lived in the Miocene and Pliocene Epochs, and may have evolved into *Stegodon*.

Stegolophodon

Source (edited): "http://en.wikipedia. org/wiki/Stegolophodon"

Stenomylus

Stenomylus is an extinct genus of miniature camelid native to North America. Its name is derived from the Greek στείνος, "narrow" and μύλος, "molar."

Restoration

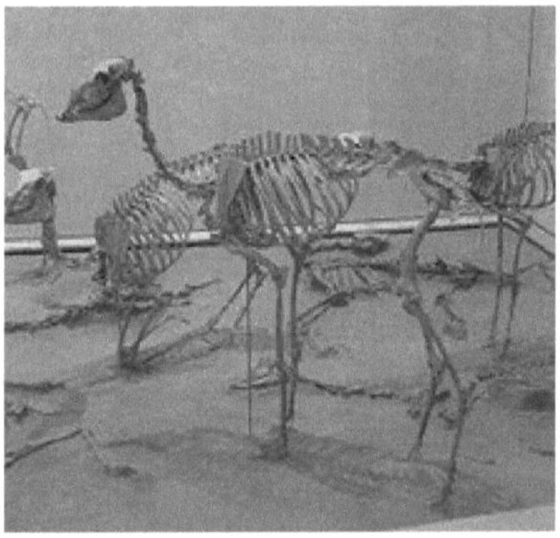

Stenomylus hitchcocki

Stenomylus was extremely diminutive compared to other ancient and modern camelids, standing only 2 feet (61 cm) tall on average. It was a slender animal with a long neck, having some resemblance to a modern gazelle. Unlike modern camelids, *Stenomylus* lacked padding on its hooves and has been compared to the modern pronghorn of North America and the gerenuk of Africa based on theories about its biomechanics.

Source (edited): "http://en.wikipedia. org/wiki/Stenomylus"

Stirtonia (monkey)

Stirtonia is an extinct genus of New World monkey from the middle and late Miocene. Its remains have been found in South America. It is classified it in the Alouattini with the howler monkeys.

Source (edited): "http://en.wikipedia. org/wiki/Stirtonia_(monkey)"

Syndyoceras

Syndyoceras is a small extinct genus of Artiodactyla, of the family Protoceratidae, endemic to western North America from the Eocene epoch (Uintan to Chadronian stage) 24.8—20.6 Ma, existing for approximately 4.2 million years.

Taxonomy

Syndyoceras was named by Barbour (1905). Its type is *Syndyoceras cooki*. It was assigned to Protoceratidae by Barbour (1905) and Carroll (1988); and to Kyptoceratini by Webb (1981), Prothero (1998), Webb et al. (2003) and Prothero and Ludtke (2007).

Morphology

Restoration

Restoration by Robert Bruce Horsfall

*Syndyoceras*s skull decorations, however, looked quite unlike those of a deer. It had two pairs of horns. The first was a V-shaped pair on the snout, fused at the base. The second pair was placed between the eyes and the ears and was curved inwards, the horns facing towards each other in a semicircular shape. Like giraffe ossicones, these protrusions were covered with skin. They were probably used for display and fighting.

In addition to the horns, *Syndoceras* also possessed tusk-like canine teeth, that it may have used to root through soil and undergrowth for food, in a similar manner to a modern musk deer. The shape of the skull also suggests that it may have had an inflated muzzle, like that of a modern saiga.

The 1.5-foot-long (0.46 m) creature closely resembled a deer, having two hooved toes. Like early horses, such as *Merychippus*, it had two vestigal outer toes on each foot, which did not touch the ground.

Source (edited): "http://en.wikipedia.org/wiki/Syndyoceras"

Synthetoceras

Synthetoceras is a large extinct genus of Artiodactyla, of the family Protoceratidae, endemic to North America from the Miocene epoch, 13.6—5.33 Ma, existing for approximately 8.27 million years.

Taxonomy

Synthetoceras was named by Stirton (1932). It is the type genus of Synthetoceratinae, Synthetoceratini. It was assigned to Protoceratidae by Stirton (1932), Thurmond and Jones (1981) and Carroll (1988); to Synthetoceratinae by Hulbert and Whitmore (2006); and to Synthetoceratini by Webb (1981), Prothero (1998), Webb et al. (2003) and Prothero and Ludtke (2007).

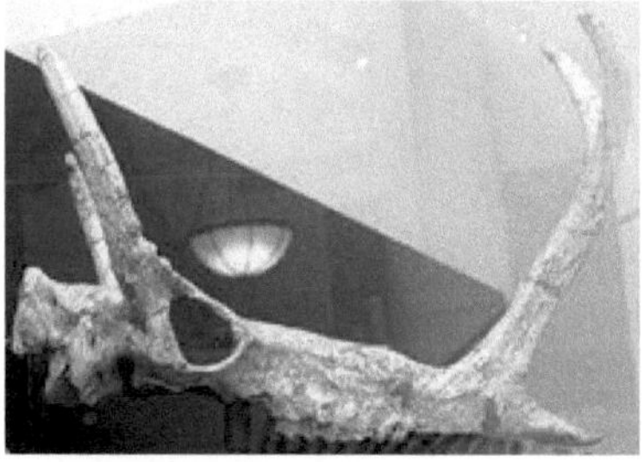

Synthetoceras tricoronatus

Morphology

With a length of 2 m (6 ft 8 in), *Synthetoceras* was the largest member of its family. It was also the last, and had what is considered to be the protoceratids' strangest set of horns. The two horns above its eyes looked fairly normal and similar to those of many modern horned mammals, but on its snout it had a bizarre, long horn with a forked tip that gave it a Y shape. Only males had this strange horn, and they probably used it in territorial fights.

Body mass

Three fossil specimens of *Poabromylus* were measured by M. Mendoza, C. M. Janis, and P. Palmqvist for body mass. The specimens were determined to weigh:

- Specimen 1: 332.4 kg (730 lb)
- Specimen 2: 228.3 kg (500 lb)

Fossil distribution

Fossils have been recovered from:

- Stage Hill II, Arikaree Group, Wildcat Hills Beds Formation, Scotts Bluff County, Nebraska
- Guernsey Site, Upper Harrison Beds

Formation, Platte County, Wyoming Source (edited): "http://en.wikipedia.org/wiki/Synthetoceras"

Synthetoceratinae

Synthetoceratinae is a subfamily of Protoceratidae (deer-like) herbivorous mammals belonging to the order Artiodactyla endemic to North America during the Miocene epoch, living 23.03—3.9 Ma, existing for approximately 19.13 million years.

Taxonomy

Synthetoceratinae was named by Frick (1937). Its type is *Synthetoceras*. It was considered monophyletic by Webb et al. (2003). It was assigned to Protoceratidae by Webb (1981), Prothero (1998), Webb et al. (2003), Hulbert and Whitmore (2006) and Prothero and Ludtke (2007).

Tribes

Synthetoceratinae contains the tribes Kyptoceratini and Synthetoceratini. Within these tribes are the genus Kyptoceras and Syndyoceras respectively.

Source (edited): "http://en.wikipedia.org/wiki/Synthetoceratinae"

Synthetoceratini

Synthetoceratini is an extinct tribe of the subfamily Synthetoceratinae within the family Protoceratidae belonging to the order Artiodactyla endemic to North America during the Miocene, living epoch 20.6—4.9 Ma, existing for approximately 15.7 million years.

Taxonomy

Synthetoceratini was named by Webb (1981). Its type is Synthetoceras. It was assigned to Synthetoceratinae by Webb (1981), Prothero (1998), Webb et al. (2003) and Prothero and Ludtke (2007).

Members

Lambdoceras, Prosynthetoceras, Synthetoceras

Source (edited): "http://en.wikipedia.org/wiki/Synthetoceratini"

Tanymykter

Tanymykter is an extinct genus of terrestrial herbivore the family Camelidae, endemic to North America during the Miocene 23.030—20.6 mya, existing for approximately 2.43 million years.

Taxonomy

Tanymykter was named by Honey and Taylor (1978). It was assigned to Camelidae by Honey and Taylor (1978) and Carroll (1988).

Morphology

Body mass

A single specimen was examined for estimated body mass by M. Mendoza, C. M. Janis, and P. Palmqvist. This specimen was estimated to weigh:

- 148.6 kg (330 lb)

Fossil distribution

Fossil distribution is somewhat widespread from Wyoming and Nebraska to Santa Barbara County, California.

Source (edited): "http://en.wikipedia.org/wiki/Tanymykter"

Tapirus polkensis

Tapirus polkensis, the **pygmy tapir**, was a small prehistoric tapir that lived in North America during the Late Miocene and early Pliocene. It is thought that *Tapirus polkensis* lived in swamps, where it would have been preyed on by ancestors of modern American crocodiles. It is smaller than any extant tapir, with an estimated mass of around 125 kilograms (280 lb). It was similar to a contemporary species of prehistoric horse known as *Parahippus* that also inhabited North American Miocene wetlands.

Source (edited): "http://en.wikipedia.org/wiki/Tapirus_polkensis"

Teleoceras

Teleoceras is an extinct genus of grazing rhinoceros that lived in North America during the Miocene epoch, which ended about 5.3 million years ago, all the way to the early Pliocene epoch.

Teleoceras had much shorter legs than modern rhinos, and a barrel chest, making its build more like that of a hippopotamus than a modern rhino. Like the hippo, it was probably also semi-aquatic. *Teleoceras* had a single small nasal horn.

Restoration

T. proterum and *Barbourofelis loveorum*

Teleoceras is the most common fossil in the Ashfall Fossil Beds of Nebraska. In fact, its remains were so numerous and concentrated that the building housing the greatest concentraion of Ashfall fossils is dubbed the "Rhino Barn". Most of the skeletons are preserved in a nearly-complete state. One extraordinary specimen includes the remains of a *Teleoceras* calf trying to suckle from its mother. This animal was featured in the episodes "Are Rhinos Dinos?" and "Dawn Of The Cats" of the Paleoworld series.

Source (edited): "http://en.wikipedia.org/wiki/Teleoceras"

Telicomys

Telicomys is an extinct genus of rodent from South America.

With a length of more than 2 metres (6.6 ft) in *T. gigantissimus*, it contains two or three of the largest rodents that ever lived, along with *Phoberomys*, *Josephoartigasia*, and the giant beaver. It is part of the same South American radiation of rodents as both *Phoberomys* and the modern capybara, which is the largest living rodent, reaching lengths of up to 1.35 metres (4.4 ft). The closest living relative to *Telicomys* is the pacarana.

Its name was made from Greek τηλικος + μυς = "a mouse [= rodent] of such a size".

Source (edited): "http://en.wikipedia.org/wiki/Telicomys"

Temnocyonines

Temnocyon are an extinct genus of carnivorous mammals belonging to the family of Amphicyonidae (Bear-dog) endemic to North America that lived during the Oligocene to Early Miocene approximately 30.3—20.6 mya existing approximately 9.7 million years.

Taxonomy

Temnocyon was named by Cope (1879). Its type is Temnocyon altigenis. It was assigned to Amphicyonidae by Cope (1879) and Carroll (1988); and to Temnocyoninae by Hunt (1998).

Morphology

A single specimen was examined for by body mass by Legendre and Roth. It was estimated to have weighed 21.3 kg (47 lb).

Fossil distribution

The first fossils are recorded in North America at Logan Butte in the John Day beds of Oregon 29–29.5 Ma., in the Sharps Formation of the Wounded Knee area, South Dakota 28–29.5 Ma, and in the Gering Formation at Wildcat Ridge, Nebraska 28.3 Ma. These early temnocyonines attained the size of coyotes or small wolves (15–30 kg) are identified by a uniquely specialized dentition. The last documented occurrences of temnocyonines are found in sediments in northwest Nebraska and southeastern Wyoming (Hunt, 2004).

Source (edited): "http://en.wikipedia.org/wiki/Temnocyonines"

Tephrocyon

Tephrocyon is an extinct genus of the Borophaginae and a small to medium sized terrestrial canine (Bear-dog) which inhabited most of North America during the Barstovian stage of the Middle Miocene 16.3—13.6 Ma, existing for approximately 2.7 million years.

Morphology

Fossil specimens of two individuals' body mass were examined by Legendre and Roth. The first specimen was estimated to weigh 12.9 kg (28 lb). The second specimen was estimated to weigh 12.1 kg (27 lb).

Fossil record

It is a rarely found genus, with fossil deposits only occurring in western Nebraska, Wyoming, New Mexico and north Florida. *Desmocyon* contains one species: *Tephrocyon rurestris*. It was first named by Merriam 1909.

Sister genera

Cormocyon, *Desmocyon*, *Euoplocyon*, *Metatomarctus*, *Microtomarctus*, *Protomarctus*, and *Psalidocyon*.

Source (edited): "http://en.wikipedia.org/wiki/Tephrocyon"

Tetralophodon

Tetralophodon
Life
Tetralophodon ("four-ridged tooth") is an extinct gomphothere genus (family Gomphotheriidae). Like typical gomphotheres, Tetralophodon had four tusks and a trunk. This genus of animals stood about ten feet tall and was a very widespread and successful proboscidean. Tetralophodon lived through the Miocene and Pliocene epoches. These animals have been found in North America, Europe, Asia, and Africa. The member species are *T. longirostris*, *T. fricki*, *T. atticus*, and *T. punjabensis*.

Description

Tetralophodon was an elephant-like animal which existed in the Miocene and Pliocene eras, approximately 2.5 million years ago. The majority of the gomphotheres went extinct at the end of the Pleistocene era, during what has been named the PT extinction. While the reason for this extinction is still debated, what is known is that these massive elephantoids under the genus Tetralophodon did not survive.

Ancestral Features

Named for its four-ridged tooth, the teeth on these animals are approximately 60 x 80 mm, about 6 times the size of a normal human tooth. These low-crowned, bunodont teeth are designed for crushing and grinding, compared with other mammals during this era that had sharp teeth used for cutting. The teeth of the tetralophodon indicate a diet of large fruits and vegetables. This diet is aided by the large size and long trunks of the elephantiods that enable these mammals to reach tall, fruit-bearing trees.

Fossil records

Most fossil records of tetralophodon are of four-ridged teeth. Their body is believed to be about 10 feet tall, with a long trunk and incisors ranging up to 6.5 feet long. These incisors are believed to be utilized as a defense mechanism. Most fossils of tetralophodon have been found in South America and Europe. The name "tetralophodon" was described in the mid-19th century with the discovery of the specialized teeth.
Source (edited): "http://en.wikipedia.org/wiki/Tetralophodon"

Thalassictis

Thalassictis is an extinct genus of terrestrial carnivore the family Hyaenidae, endemic to Asia and Southern Europe during the Early Miocene subepoch, 8.7—5.3 mya, existing for approximately 3.4 million years.

Taxonomy

Thalassictis was named by Nordmann (1850) [in Gervais]. Its type is *Thalassictis robusta*. It was assigned to Hyaenidae by Kurtén (1982) and Flynn (1998).
Source (edited): "http://en.wikipedia.org/wiki/Thalassictis"

Thalassocnus

Closeup of *Thalassocnus* skull, Muséum national d'histoire naturelle, Paris

Thalassocnus is an extinct genus of semi-aquatic or aquatic marine sloth from the Miocene and Pliocene of South America. Fossils found to date have been from the coast of Peru. They were apparently grazers of sea grass and seaweed. Over time, they apparently shifted from a preference for feeding in shallow water to a preference for deeper waters. They may have used their powerful claws to anchor themselves to the sea floor to facilitate feeding, similar to the behavior of the marine iguana.
Source (edited): "http://en.wikipedia.org/wiki/Thalassocnus"

Thalassoleon

Thalassoleon is an extinct genus of large fur seal. *Thalassoleon* inhabited the Northern Pacific ocean in latest Miocene and early Pliocene time. Fossils of *T. mexicanus* are known from Baja California and southern California. *T.*

macnallyae is known from central California, and *T. inouei* (which may be a synonym of *T. macnallyae*) is known from Japan. *Thalassoleon* could be the ancestor of the modern Northern Fur Seal.

Source (edited): "http://en.wikipedia.org/wiki/Thalassoleon"

Thaumastocyonini

Thaumastocyonini is an extinct tribe of large, carnivorous mammals (bone-crushers) known as bear dogs, of the family Amphicyonidae, subfamily Amphicyoninae, which inhabited Europe during the Miocene living from ~16.9—7.25 Ma and existed for approximately 9.65 million years.

Taxonomy

Thaumastocyonini was named by Ginsburg (1977). It was assigned to Amphicyoninae by Ginsburg (1977).

Fossil distribution

- Bèni Mellal, Morocco ~16.0—11.6 Ma.
- En Pejouan, France ~15.7—7.25 Ma.
- Pontigne, France ~11.61—7.25 Ma.
- Baigneaux-en-Beauce, France ~16.9—16.0 Ma.

Source (edited): "http://en.wikipedia.org/wiki/Thaumastocyonini"

Theosodon

Theosodon is an extinct genus of litoptern mammal from the early Miocene of South America.

Restoration of *Theosodon garretorum* and *Borhyaena tuberata*

Theosodon bore a superficial resemblance the modern guanaco, and was around 2 metres (6.6 ft) in length. It had a long neck and tapir-like, three-toed feet. Judging from the position of its nostrils, *Theosodon* had a short trunk, though shorter than its other relatives, such as *Macrauchenia*.

Theosodon had a full set of 44 teeth, and its lower mandible was very thin and large.

Source (edited): "http://en.wikipedia.org/wiki/Theosodon"

Thylacinus potens

Thylacinus potens (*"powerful thylacine"*) was the largest species of the family Thylacinidae, known from a single poorly preserved fossil discovered by Michael O. Woodburne in 1967 in a Late Miocene locality near Alice Springs, Northern Territory. It preceded the modern thylacine by 4–6 million years, and was 5% bigger, was more robust and had a shorter, broader skull. Its size is estimated to be similar to that of a grey wolf; the head and body together were around 5 feet long, and its teeth were less adapted for shearing compared to those of the modern thylacine.

Source (edited): "http://en.wikipedia.org/wiki/Thylacinus_potens"

Thylacosmilus

Thylacosmilus ("pouch sabre") was a genus of sabre-toothed metatherian predators that first appeared during the Miocene. Remains of the animal have been found in parts of South America, primarily Argentina. Though *Thylacosmilus* is one of several predatory mammal genera typically called "sabre-toothed cats", it was not a felid but a sparassodont, a group closely related to marsupials, and only superficially resembled other sabre-toothed mammals due to convergent evolution.

Description

Thylacosmilus had long, sabre-like upper canines and short, blunt, peg-like lower canines. The incisors were missing altogether and the other teeth were severely reduced, but, as distinct from machairodonts, their number was complete. It is estimated to have weighed around 150 kilograms (330 lb).

Thylacosmilus' sabre-teeth kept growing throughout its life, unlike those of true sabre-toothed felines. It also had a pair of elongated, scabbard-like flanges growing from the lower jaw, which protected the sabre-teeth when it closed its mouth. The cervical vertebrae were very strong and to some extent resembled the vertebrae of machairodonts.

Thylacosmilus went extinct during the Pliocene, a timeframe closely corresponding to the arrival of the sabre-toothed cat *Smilodon* from North America in the Great American Interchange. As such out-competition from the anatomically similar *Smilodon* appears to have driven the extinction, though this is unusual among other South American extinctions during the period.

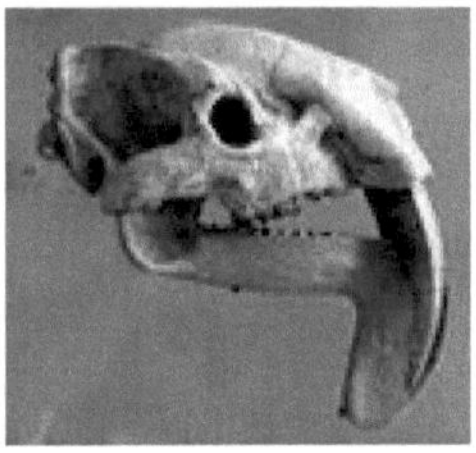

Thylacosmilus atrox skull, American Museum of Natural History.

An artist's interpretation of *Thylacosmilus*.

Source (edited): "http://en.wikipedia.org/wiki/Thylacosmilus"

Ticholeptus

Ticholeptus is an extinct genus of oreodont endemic to North America during the Early Miocene-Middle Miocene epochs (20.6—13.6 mya), existing for approximately 7 million years.

Taxonomy

Ticholeptus was named by Cope (1878). Its type is *Ticholeptus zygomaticus*. It was synonymized subjectively with *Merychyus* by Cope (1884), Scott (1890) and Scott (1893). It was assigned to Merycoidodontidae by Cope (1878), Thorpe (1937), Schultz and Falkenbach (1941), Kelly and Lander (1988) and Lander (1998)...

Morphology

Two specimens were examined by M. Mendoza for body mass. Specimen one was estimated to have a weight of 116 kg (255.7 lbs). Specimen two was estimated to have a weight of 154.9 kg (341.4 lbs).

Fossil distribution

Fossils have been uncovered throughout the U.S. from Gadsden County, Florida to Jefferson County, Oregon as well as Ventura, California, numerous sites in Nebraska, Nevada, and Montana.

Species

T. zygomaticus (syn. *Merycochoerus obliquidens, Poatrephes paludicola, T. hypsodus, T. rileyi, Ustatochoerus schrammi*).

Source (edited): "http://en.wikipedia.org/wiki/Ticholeptus"

Titanotylopus

Titanotylopus (synonym ***Gigantocamelus***) is an extinct genus of terrestrial herbivore the family Camelidae, endemic to North America from the Miocene through Pleistocene 10.3 mya—300,000 years ago, existing for approximately 10 million years.

Its name is derived from the Greek words Τιτάν, τύλος and πούς — "Titan", "knob" and "foot"; thus, "giant knobby-foot."

Paleobiology

Titanotylopus is distinguished from other early large camelids by its large upper canines and other distinguishing dental characteristics, and absence of lacrimal vacuities in the skull. Unlike the smaller, contemporaneous *Camelops*, *Titanotylopus* had relatively broad second phalanges, suggesting that it had true padded "cameltoes," like modern camels..

The species *Titanotylopus spatulus* was characterized by broad, spatula-like incisors. It has been found at Grand View, Red Light, Hudspeth County, Texas, Donnelly Ranch, White Rock, Mullen II (Kansas), Sandahl Local Fauna (Nebraska) and Vallecito Creek, Colorado, 111 Ranch, Arizona in North America.

Appearance

Titanotylopus possessed long and massive limbs, a comparatively small braincase, and a convex slope between the eyes. Its average height was 3.5 metres (11 ft).

Like modern camels, it possessed a hump for fat storage; evidence for this is provided by the long neural spines on its thoracic vertebrae .

Alternate classification

While some authors have considered *Gigantocamelus* and *Titanotylopus* to be congeneric, others have maintained them separately. Voorhies and Corner, based on previously unreported material, documented that the two are indeed worthy of separate generic status. Harrison (1985) followed Voorhies and Corner in advocating the use of *Titanotylopus* for only *T. nebraskensis*, based on a lower jaw, and *Gigantocamelus* for *G. spatula*, which includes *G. fricki*. There is a clear difference between the proximal phalanx of specimens assigned to *Gigantocamelus* and to *Titanotylopus*, based on skeletons associated with skull material.

Source (edited): "http://en.wikipedia.

org/wiki/Titanotylopus"

Tomarctus

Tomarctus is a canine of the extinct subfamily Borophaginae which inhabited most of North America during the late Early Miocene to the Early Barstovian age of the Middle Miocene (23—16 mya). *Tomarctus* existed for approximately 6.83 million years.

Tomarctus evolved from the earlier genus *Nothocyon*. This animal shared a period of time and ecology with a variety of other bear dogs like the giant mustelid genus of bone-crushing canidae, *Cynarctoides*. As the bear dogs and giant mustelids became extinct, *Tomarctus* further radiated to fill a line of dogs which filled the hyena-like fruit eating and bone-crushing niches.

Species

The genus currently contains two accepted species, *Tomarctus brevirostris* and *Tomarctus hippophaga*.

Tomarctus brevirostris, synomymous with *Aelurodon francisi*, was named by Edward Drinker Cope in 1873. Fossil specimens have been found as far south as Panama, east to Plum Point, Maryland, west to California, and north to Montana.

Tomarctus hippophaga was first described by Matthew and Cook in 1909 from the Trojan Quarry, Olcott Formation, Nebraska. Specimens have since been found as far west as California and as far north as the Montana/Alberta, Canada line.

Source (edited): "http://en.wikipedia.org/wiki/Tomarctus"

Toxodonta

Toxodonta is a suborder of the extinct mammalian order Notoungulata. The group is named after *Toxodon*, the first example of the group to be discovered. The suborder includes the largest known notoungulates, members of the family Toxodontidae. Toxodontids first appeared during the Oligocene in the form of *Proadinotherium*. Another well-known toxodontid is *Nesodon*, a medium-sized Miocene toxodontid descended from *Proadinotherium* which had converted its second upper incisors into tusks. *Toxodon* itself evolved during the Pliocene alongside the related *Trigodon*, an equally large mammal which possessed a horn projecting from its forehead, in the same way as a modern rhinoceros. *Toxodon* was a huge herbivore (about the size of a modern black rhinoceros) with four toes on each hoof.

Closely related to the toxodontids are two additional families: the Leontiniidae of the Eocene through Miocene and the Notohippidae. The last surviving notohippid was *Colpodon* from the late Oligocene, which was medium-sized but heavily built. Notohippids possessed molars that were superficially like those of modern horses.

Source (edited): "http://en.wikipedia.org/wiki/Toxodonta"

Toxodontidae

Xotodon sp. skull at the Museum für Naturkunde, Berlin

Toxodontidae is an extinct family of notoungulate mammals known from the Oligocene through the Pleistocene of South America, with one genus, *Mixotoxodon*, also known from the Pleistocene of Central America. They somewhat resembled rhinoceroses, and had teeth with high crowns and open roots, suggesting that they often fed on tough pampas grass. However, isotopic analyses have led to the conclusion that the most recent forms were grazing and browsing generalists.

Source (edited): "http://en.wikipedia.org/wiki/Toxodontidae"

Tremarctini

Tremarctini, the **short-faced bears**, is a tribe belonging to the family Ursidae (bears) endemic to North America and Europe during the Miocene to Holocene living from ~13.6 Ma—present, existing for approximately 13.6 million years.

Tremarctini was named by Frick (1926). Its type is *Plionarctos edensis*.

It was assigned to Ursidae by Frick (1926) and Carroll (1988); and to Tremarctinae by Hunt (1998). This tribe includes four genera: *Arctodus*, *Arctotherium*, *Plionarctos* and *Tremarctos*, with the latter containing both extinct and extant species (e.g., Florida Cave Bear and Spectacled Bear).

Fossil distribution

Sites and specimen ages:

- San Josecito Cave, Mexico *(Tremarctos floridanus)* ~100,000—11,000 years ago
- Bear Tooth Slide, Brown County, Nebraska *(Plionarctos)* ~13.6—4.9 Ma.
- Taunton site, Adams County, Washington *(Plionarctos harroldorum)* ~4.9—1.8 Ma.
- Pipe Creek Sinkhole, Grant County, Indiana *(Plionarctos edensis)* ~10.3—1.8 Ma.

Source (edited): "http://en.wikipedia.org/wiki/Tremarctini"

Triceromeryx

Triceromeryx is an extinct genus of Artiodactyla, of the family Palaeomerycidae, endemic to Europe from the early Miocene epoch, 22.4—20.0 Ma, existing for approximately 2.4 million years.

It was similar to *Ampelomeryx*, a herbivore.

Taxonomy

Triceromeryx was named by Villalta Commela et al. (1946). It was assigned to Giraffidae by Carroll (1988); and to Palaeomerycinae by Prothero and Liter (2007).

Fossil distribution

- Cetina de Aragon, Zaragoza, Spain

Source (edited): "http://en.wikipedia.org/wiki/Triceromeryx"

Trigodon

Trigodon gaudryi is an extinct species of the family Toxodontidae, a large bodied notoungulate which inhabited South America during the Miocene living from 11.61—7.25 Ma and existed for approximately 4.36 million years.

It bore a strong, albeit superficial resemblance to a rhinoceros, in that it had a horn on its forehead.

Fossil distribution

- Solimoes Formation, Acre Province, Brazil, eastern slope of Andes Mountains.

Source (edited): "http://en.wikipedia.org/wiki/Trigodon"

Tupaia miocenica

Tupaia miocenica is a fossil treeshrew from the Miocene of Thailand. Known only from a single tooth, an upper second molar, it is among the few known fossil treeshrews. With a length of 3.57 mm, the tooth is large for a treeshrew. At the back lingual corner (the side of the tongue), the tooth shows a small cusp, the hypocone, that is separated from the protocone in front of it by a narrow valley. The condition of the hypocone distinguishes this species from various other treeshrews. In addition, the presence of a well-developed but simple mesostyle (a small cuspule) is distinctive.

Taxonomy

Tupaia miocenica was described in 1997 by French paleontologists Pierre Mein and Léonard Ginsburg in a report on the fossil mammals of Li Mae Long, a Miocene site in Thailand. The animal is known from a single tooth, which according to Mein and Ginsburg's comparisons most closely resembles the living treeshrews of the genus *Tupaia*. The specific name, *miocenica*, refers to the animal's occurrence during the Miocene. Only a few other fossil treeshrews are known, and *T. miocenica* is the only fossil species from Thailand as well as the only Miocene representative of *Tupaia*.

Description

The single known tooth, a worn left second upper molar (M3) known as T Li 175, is large for a treeshrew, with a length of 3.57 mm and width of 4.79 mm. It is dilambdodont (with a W-shaped chewing surface) and the labial surface (the side of the cheeks) is concave and bears a cingular crest. A well-developed small cusp, a mesostyle, is present on the labial side. The presence of the mesostyle distinguishes the animal from *Ptilocercus*, the only member of the treeshrew family Ptilocercidae. The lingual (tongue) side is narrow. A large cusp, the protocone, is on the front lingual corner. Two crests descend from it; one reaches the paracone on the front labial side and another approaches the metacone to the back of the tooth. A smaller cusp, a hypocone, is on the back lingual corner. This feature distinguishes *T. miocenica* both from the treeshrew genera *Dendrogale*, *Prodendrogale*, and *Palaeotupaia*, which lack the cusp entirely, and from *Anathana* and *Urogale*, which have a large hypocone. In *T. miocenica* the back and lingual sides of the hypocone are straight and form a right angle with each other. The tooth most closely resembles *Tupaia* species with a simple mesostyle, such as the common treeshrew (*Tupaia glis*). However, this species is smaller and unlike in *T. miocenica*, the hypocone is not isolated from the protocone by a narrow valley.

Range and ecology

Li Mae Long, the collection site of *T. miocenica*, is dated to the latest Early Miocene, corresponding to the European zone MN 4, around 18 mya. It is in the Thai province of Lamphun. The fossil fauna encompasses 34 species of mammals, including the tarsier *Tarsius thailandica* and the slow loris ? *Nycticebus linglom*. Mein and Ginsburg conclude that the fauna represents a tropical forest environment close to a shallow lake.

Source (edited): "http://en.wikipedia.org/wiki/Tupaia_miocenica"

Tylocephalonyx

Tylocephalonyx is an extinct dome-headed chalicothere from the Miocene in North America. It may have used its "dome" in the same way as the pachycephalosaurs. *Tylocephalonyx* is related to the modern horse, rhino, and tapir.

Source (edited): "http://en.wikipedia.org/wiki/Tylocephalonyx"

Ursavus

Ursavus is an extinct genus of mammals of the family Ursidae (bears) that existed in North America, Europe, Africa, and Asia during the Miocene, living from ~23—5.3 Ma, existing for approximately 17.7 million years. It apparently dispersed from Asia into North America about 20 Ma, becoming the earliest member of the subfamily Ursinae in the New World. Qiu points out that if a questionable 29 million-year-old specimen of Ursavus reported in North America is validated, Ursavus may have evolved in North America and dispersed westward into Asia. The higher number of fossils in Europe grading toward eastern Asia make the westward dispersal unlikely.

Ursavus was named by Schlosser (1899). It was assigned to Ursidae by Schlosser (1899) and R. L. Carroll (1988); and to Ursavini by R.M. Hunt (1998) and Jin et al. (2007).

Fossil distribution

Sites (not complete) and specimen ages:

- Pawnee Buttes Site, Weld County, Colorado *(U. pawniensis)* ~23.03—5.3 Ma.
- Pasalar site, Bursa, Turkey *(U. primaevus)* ~16—13.7 Ma.
- Baigneaux-en-Beauc, Alsac, France *(U. brevirhinus)* ~16.9—16.0 Ma.
- Hambach mine horizon 6C, Germany *(U. elmensis* ~16.9—13.7 Ma.
- Yost Farm Site, Saskatchewan, Canada *(U. primaevus)* ~16.3—13.6 Ma.
- Myers Farm Site, Valentine Formation, Webster County, Nebraska *(U. brevirhinus)* ~16.3—13.6 Ma.
- Lufeng site, Yunnan, China *(U. depereti)* ~9—5.3 Ma.

Source (edited): "http://en.wikipedia.org/wiki/Ursavus"

Vanderhoofius

Vanderhoofius is a extinct genus of herbivorous marine mammal of the family *Desmostylidae* living from the Middle Miocene subepoch (15.9 mya—11.6 Mya) and in existence for approximately 4 million years.

Morphology

Vanderhoofius was a large, hippopotamus-like creature. It had a short tail and powerful legs with four hooves. Both the creature's jaws were elongated and sported forward-facing tusks, which were elongated canines and incisors.

Species

V. coalingensis

Sister genera

Behemotops, Cornwallius, Desmostylus, Kronokotherium, Paleoparadoxia,

Fossil distribution

Fossils have been discovered only in California

Source (edited): "http://en.wikipedia.org/wiki/Vanderhoofius"

Veratalpa

Veratalpa lugdunensiana is a fossil mammal from the Miocene of France. Known from a single astragalus (a foot-bone), the species was assigned to its own genus, *Veratalpa*, by Florentino Ameghino in 1905. He placed it in Talpidae, the family of the moles, but in 1974, John Howard Hutchison argued that the astragalus was not talpid and more likely came from a rodent. The astragalus is about 4.5 mm long, broad for a talpid, and has the head oriented farther from the axis of the foot than in talpids.

Taxonomy

Argentine naturalist Florentino Ameghino described *Veratalpa* in an overview of the astragali from the middle Miocene of Vieux Collonges in France. The astragalus is a bone of the foot that is part of the ankle joint. He listed several species of the family Talpidae (moles and related species) from Vieux Collonges, including "espèce C" ("species C"), which he named as a new genus and species, *Veratalpa lugdunensiana*, in a footnote. In a 1906 review of Ameghino's paper, Édouard Louis Trouessart affirmed that *Veratalpa* probably represented a new genus of mole, but noted that the specific name *lugdunensiana* would have been more correctly written "*lugdunensis*". According to Trouessart, the suffix *-ana* is appropriate for names that reference persons, but not for those that refer to places, such as this name, which is derived from *Lugdunum* (the Latin name for Lyon).

In a 1974 review of Miocene European talpids, John Howard Hutchison wrote that the astragalus of *Veratalpa* lacked any features that would ally it with talpids and commented that it was most likely a rodent. In their 1997 *Classification of Mammals*, Malcolm McKenna and Susan Bell listed *Veratalpa* as a member of Placentalia of uncertain affinities.

Description

The astragalus of *Veratalpa* is the largest among those from Vieux Collonges that Ameghino assigned to Talpidae. Although at 4.5 mm it is about as long as his "species A", it is broader, and Hutchison noted the broadness as one of the characters that argue against classification of *Veratalpa* in Talpidae. Like living moles, it has a broad, flat, and short head, but it forms a noticeably small angle with the body—in actual moles, the head is more axially oriented (i.e., towards the central axis of the foot). The surface of the head that contacts the navicula is less rounded than in moles. The body is low and nearly square and has a diameter of about 3 mm. The trochlea—a surface on the body of the bone that articulates with the tibia (lower leg bone)—is not large and pulley-like, as in talpids. There is a small perforation on the lower side of the body. This perforation is larger in Ameghino's other supposed talpids, and Trouessart suggested on the basis of this feature that the internal parts of the toes were reduced in *Veratalpa*.

Distribution

Veratalpa is known only from the locality of Vieux Collonges near Lyon in southeastern France; Ameghino knew this locality as "Mont-Ceindre". This rich fissure filling locality has yielded thousands of fossils and is currently dated to the early-middle Miocene boundary, around 17 million years ago (MN 4/5 in the MN zonation). As *Veratalpa* is known from a single astragalus, Ameghino considered it to be rare. He distinguished six talpid species among the astragali, but according to Hutchison only Ameghino's species F (which was assigned to Talpidae with a query) is really a talpid.

Source (edited): "http://en.wikipedia.org/wiki/Veratalpa"

Victoriapithecus macinnesi

Victoriapithecus macinnesi was a primate. It was described from a single fossil specimen, the oldest Old World monkey skull fossil. It was discovered near Lake Victoria in Kenya and is from the middle Miocene and was closely related to the two or three extinct *Prohylobates* species.

Morphology

Victoriapithecus macinnesi had a dental formula of 2:1:2:3 on both the upper and lower jaws. This species had lower molars that are bilophodont with low cusps. The canines of *Victoriapithecus macinnesi* show sexual dimorphism and the mandible of this species is relatively deep compared to other Old World monkeys. On the forelimbs, the distal end of the humerus shows a narrow articulation and a deep ulnar notch. *Victoriapithecus macinnesi* had an average body mass of around 7.0 kilograms.

Range and locomotion

Victoriapithecus macinnesi lived in Africa and this species was likely a quadruped due to the postcranial remains.

Source (edited): "http://en.wikipedia.org/wiki/Victoriapithecus_macinnesi"

Viverra

Viverra, is a genus of civet commonly found in Southeast Asia.

Species

- Malabar Large-spotted Civet (*Viverra civettina*)
- Large-spotted Civet (*Viverra megaspila*)
- Malayan Civet (*Viverra tangalunga*)
- Large Indian Civet (*Viverra zibetha*)

Source (edited): "http://en.wikipedia.org/wiki/Viverra"

Vulpes riffautae

Vulpes riffautae is an extinct species of fox from the late Miocene of Chad (approximately 7 ma). Fossils of *V. rif-*

fautae potentially represent the earliest record of the dog family (Canidae) in the Old World. *V. riffautae* was intermediate in size between the Rüppell's Fox (*Vulpes rueppellii*) and Fennec Fox (*V. zerda*).

Source (edited): "http://en.wikipedia.org/wiki/Vulpes_riffautae"

Wakaleo

Wakaleo (indigenous Australian *waka*, "little", "small", and Latin *leo*, "lion"), was a genus of medium-sized thylacoleonids that lived in Australia in the early to late Miocene. It was approximately 2.5 ft (80 cm) long, or the size of a dog. Although much smaller than its close relative, the Marsupial Lion *(Thylacoleo carnifex)*, *Wakaleo* would have been a successful hunter in its time. It had teeth specially designed for cutting and stabbing.

Species

- *Wakaleo alcootaensis* was found in the Miocene Waite Formation in the Northern Territory in 1974. It was slightly larger than the other two species.
- *Wakaleo oldfieldi* was found by a group of scientists working in the Miocene Wipijiri Formation in southern Australia in 1971. They found a nearly complete left dentry which included a few well-preserved teeth.
- *Wakaleo vanderleuri* was first found by some field workers in 1967 in the Miocene Camfield Beds in the Northern Territory. Many more specimens have been found since then.

Source (edited): "http://en.wikipedia.org/wiki/Wakaleo"

Wakaleo oldfieldi

Wakaleo oldfieldi is an extinct species of marsupial lion found in the tertiary deposits of South Australia. There are three unfused molar teeth instead of two fused molars as is the case with the Pleistocene *Thylacoleo carnifex.*

As with *T. carnifex*, this species is presumed to used its maxillary (upper) teeth to hold its food and sharpen the mandibular teeth, the later were also used in slicing and stabbing during eating. The premolars also had a crescent shaped circumference for slicing.

Source (edited): "http://en.wikipedia.org/wiki/Wakaleo_oldfieldi"

Wakaleo vanderleuri

Wakaleo vanderleuri is a species of "marsupial lion" that lived in Australia during the Miocene (approximately 16 to 10 million years ago). It is not closely related to true cats, as being a marsupial its closest living relatives are wombats and the Koala. This diprotodont is hypothesized to have been an omnivore, eating significant amounts of meat, perhaps more like a bear in diet and habit than like an actual lion. It lacked large canine-like fangs, producing some controversy regarding just how much it actually preyed on living animals. The meat-eating hypothesis is based largely upon the morphology of its molar teeth. The species is believed to have been able to hunt by leaping from trees onto its prey.

The type species fossils were originally found in the Bollock Creek area.

Source (edited): "http://en.wikipedia.org/wiki/Wakaleo_vanderleuri"

Warendja

Warendja is an extinct genus of marsupial.

Source (edited): "http://en.wikipedia.org/wiki/Warendja"

Yarala

Yarala is a genus of fossil mammals that resemble contemporary bandicoots. The superfamily Yaraloidea and family Yaralidae were created following the discovery of the type species *Yarala burchfieldi* in 1995, on the basis that it lacks synapomorphies that unite all other peramelemorphian taxa.

A second species was described in 2006, which is suggested to be ancestral to *Y. burchfieldi.*

Source (edited): "http://en.wikipedia.org/wiki/Yarala"

Ysengrinia

Ysengrinia is an extinct genus of large, mostly carnivorous bone-crushing mammals known as bear dogs, of the family Amphicyonidae endemic to Europe, Africa, and North America during the Miocene living from 22.4—20 Ma and existed for approximately 2.4 million years.

Taxonomy

Ysengrinia was named by Leidy (1853). was named by Ginsburg (1965). It was assigned to Amphicyoninae by Hunt (1998); and to Amphicyonidae by Ginsburg (1965), Carroll (1988) and Hunt (2002).

Morphology

A single specimen was examined by Legendre and Roth for body mass and was estimated to weigh 71.6 kg (160 lb).

Source (edited): "http://en.wikipedia.org/wiki/Ysengrinia"

Yumaceras

Yumaceras is an extinct genus of horse-like Artiodactyla, of the family Palaeomerycidae, endemic to North America, Europe and Asia from the Miocene epoch, 13.6—5.33 Ma, existing for approximately 8.27 million years.

Taxonomy

Yumaceras was named by Frick (1937). It was synonymized subjectively with Pediomeryx by Savage (1941) and Janis and Manning (1998); it was reranked as Pediomeryx (Yumaceras) by Webb (1983); it was synonymized subjectively with Cranioceras by Tedford et al. (1987). It was assigned to Pediomeryx by Webb (1983); and to Cranioceratini by Prothero and Liter (2007).

Fossil distribution

- Norris Canyon, Contra Costa County, California
- Cambridge Site, Frontier County, Nebraska
- Haile V/XIXA, Alachua County, Florida

Source (edited): "http://en.wikipedia.org/wiki/Yumaceras"

Zygolophodon

Zygolophodon is an extinct genus of African, Asian, North American, and European mastodon that lived during the Miocene and Pliocene epochs. It may have evolved from *Tetralophodon*. While collecting fossils in the Clarno Formation of Oregon during 1941, noted paleobotanists Alonzo W. Hancock and Chester A. Arnold recovered the most complete *Zygolophodon* skull known at the time.

Source (edited): "http://en.wikipedia.org/wiki/Zygolophodon"

Zygophyseter

Zygophyseter

Zygophyseter varolai is an extinct cetacean, similar to the sperm whale. The common name of "killer sperm whale" refers both to its relation to modern day sperm whales, and to its similarity in size to and its "probable similar feeding adaptation to the extant delphinid killer whale (*Orcinus orca*)".

The description for *Zygophyseter varolai*, published in 2006, is the only species in the genus ***Zygophyseter***. This was based on a fossil, an almost complete skeleton, which dates from the Tortonian Age of the Late Miocene Epoch of Southern Italy. It has a very long zygomatic process, probably for housing the spermaceti organ. *Z. varolai* has a large body size with large teeth present in both jaws making it capable of feeding on large prey.

Related species

- *Aulophyseter*
- *Orycterocetus*
- *Acrophyseter*
- *Brygmophyseter*
- *Livyatan melvillei*

Source (edited): "http://en.wikipedia.org/wiki/Zygophyseter"